# EASY GUIDE TO MATH & ALGEBRA

FALL RIVER PRESS
New York

FALL RIVER PRESS

New York

An Imprint of Sterling Publishing
1166 Avenue of the Americas
New York, NY 10036

FALL RIVER PRESS and the distinctive Fall River Press logo are registered trademarks of Barnes & Noble, Inc.



Text by Anna Medvedovsky and Kathy Furgang

ISBN 978-1-4351-6135-1

For information about custom editions, special sales, and premium and corporate purchases, please contact Sterling Special Sales at 800-805-5489 or specialsales@sterlingpublishing.com.

This special edition was created exclusively for BTMS.

Manufactured in the United States of America

2 4 6 8 10 9 7 5 3 1

www.sterlingpublishing.com

# CONTENTS

# EASY GUIDE TO MATH

# CONTENTS

# A FEW ENCOURAGING WORDS

## SUCK AT MATH?

This book is for people who, for whatever reason—bad teachers, bad attitude, bad brains—happen to suck at math. Hey, it happens. It's nothing to mince words about. Are you one of those people? Then read on.

My name is Shanon Fletcher. Yes, Shanon with one *n*. My parents can't spell. They're also up to something suspicious and weird. But more on that later. Anyway, I am your no-fear math tutor. Unlike you, I do not suck at math. And, lucky for you, I am here to help.

We're going to start from the very beginning. No prerequisites. The chapters go in order, but feel free to skim things you already know. All the chapters have fascinating titles like **Factoring** and **Fractions**. I take these chapters and chew them further, spitting out bite-size math morsels as I go along. Like a partially digested Baby Ruth, these worked-over math mouthfuls should be easy for you to swallow. Personally, I also find them yummy. But I can understand how *some* people might find a chewed-up mess of chocolate, caramel, and decimals a bit disgusting. That's fine. I don't offend easily.

To keep you from getting too bored while we talk math, I'm going to tell you a little story as we go along. This story is 100 percent true.* It's full of scandal, violence, rock 'n' roll, duplicity, pathetic pining, and those misbehaving parents I already mentioned. This story is my gift to you.

Your gift to me? Answer the review questions at the end of each chapter. They're the things called **Your Turn**. With any luck, this will be the last time you'll have to learn how to add fractions. Ever. If I were twenty-one, I'd drink to that.

---

*To the best of my recollection.

# EASY GUIDE TO MATH

# 1

# CHAPTER 1
## NUMBERS

### WAITING FOR BILLY IDOL

So let's dive right in. I live in Maplewood. It's a suburb of Chicago. I'm a junior at Maplewood High School. I'm seventeen years old. And I may have neglected to tell you one thing in the intro. You may have figured it out, but don't think I came out and just said it: I'm kind of a freak.

"She really is."

Thanks, Rebecca. That's my friend Rebecca chiming in. She's sitting here next to me at the café tables outside the Finer Diner. We're having an after-school snack.

"Hi."

Anyway, like I said, and like Becca confirmed, I'm strange. You want proof? I love chewing on my toenails. Especially late at night, when I'm tired. Last Saturday night, when Rebecca was sleeping over at my house, I accidentally bit her foot because I was really tired and couldn't see very well. She was pretty pissed.

"I got over it."

And here's more proof of my weirdness. I'm taking a remedial math class after school even though—as I already told you—I *kill* at math.

"She really does. I ask her math stuff all the time. It's scary how much she knows."

Again, thanks, Rebecca, but I'm trying to tell a story here. So why am I taking a beginner's math class if I'm already a math genius?

"Two words: Kyle Thomas."

Correct. Kyle Thomas is the lead singer of Johnny London—and the dreamiest boy to hit our shores since Billy Idol.

"Shanon, you are *so* retro! For anyone under seventy reading this out there, Billy Idol was a pop-punk superstar in, like, 1985. Like, before you were born."

Thanks for the history lesson, Rebecca. As I was saying, Kyle Thomas looks just like Billy, circa Billy's 1986 *Rebel Yell* tour. *Plus* Kyle's also British! His accent makes me want to . . . well, I won't get into exactly what it makes me want to do. Anyway, he's in this after-school class for the math-challenged.

"So am I. I'm tough, but numbers scare me."

"Rebecca, will you stop interrupting?" With some people it's me, me, me. Anyway, getting back to me, I've never been in a class with Kyle because Maplewood High is the size of a small country. So taking this class was my only way to get close enough to him for him to realize what a wonderful girlfriend I'd make.

"You forgot to tell them how you lied to get into the class."

Oh yeah. I knew my mom wouldn't pay for a math class that I didn't need, so I told her it was a precalculus class. That's not technically a lie—any math class you take *before* calculus is *pre*-calculus, right? Anyway, Mom forked over the dough pronto.

"That was so wrong, Shanon. You know your mom is barely hanging on mentally with your dad running around with his beautiful blond secretary! If my dad did that, believe me, my mom'd have him strapped to a gurney and subjected to one of those intense 24/7 interrogations involving truth serum and an IV drip."

"What are you talking about?"

"Oh, nothing . . ."

Anyway, I know Rebecca's right. It was very selfish of me. I do feel terrible. But I couldn't help it. I want Kyle Thomas *so* much. Want proof? Rebecca and I have planted ourselves outside the Finer Diner for one reason only.

"We're waiting for Kyle to walk by on his way to class. This is so pathetic."

But wait! I'm also helping Rebecca understand all about numbers and digits and place value and other things like that. Lest she forget how clueless she is about math.

"She's right. I am."

Thank you.

And if Kyle happens to walk by while I'm helping Rebecca . . . well, it would be simply rude not to say hi.

# NUMBERS

People use many different words when talking about numbers. It's a bit arbitrary and way dull, but let's go over this stuff so we're all on the same page.

## NATURAL NUMBERS

The **natural numbers** are the numbers you count with. Zero is not included.

$$1, 2, 3, 4, 5, 6, 7, \ldots, 25, 26, 27, \ldots, 149, 150, \ldots$$

Essentially, a natural number is any number that could be used to count discrete physical objects or ideas—7 *Rebel Yell* records or 48 cents or 365 days.

The natural numbers keep on increasing without end. There is no largest natural number.

## WHOLE NUMBERS

The **whole numbers** are the numbers you count with and zero.

$$0, 1, 2, 3, 4, 5, 6, 7, 8, 9, \ldots$$

Every natural number is a whole number. The only whole number that's not a natural number is $0$.

Is it silly to have names for two groups of numbers that are almost the same? Sort of. Fortunately you don't really have to worry about it. The difference between natural numbers and whole numbers (which is just $0$, remember?) almost never comes up.

The first six chapters—on addition, subtraction, multiplication, division, and factoring—will deal with whole numbers only. We'll sometimes plot them on a **number line**, which looks like this:

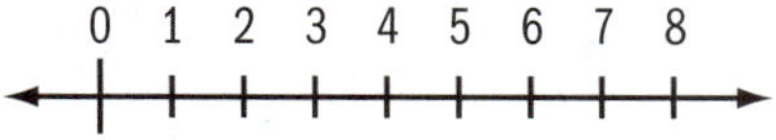


Zero is usually on the left, and the numbers increase left to right.

## POSITIVE AND NEGATIVE NUMBERS

The numbers to the right of zero on the number line are called **positive numbers**. All the natural numbers are positive.

We can also plot numbers to the left of zero. These are called **negative numbers**. To mark their negativity, we write them with a **minus sign** (–) in front, like so:

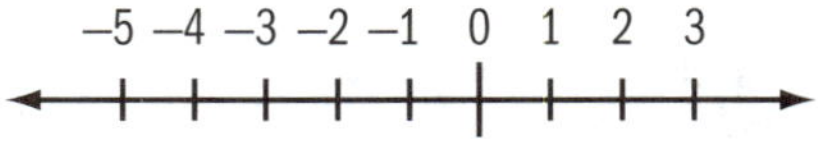


The numbers still increase from left to right. So the number $-2$ is smaller than $-1$, the number $-15$ is smaller than $-2$, and all negative numbers are smaller than $0$ and than all the positive numbers.

Each positive number has a corresponding negative number and vice versa. So the negative number $-3$ matches up with the positive number $3$. Similarly, $11$ matches up with $-11$. A pair of numbers that correspond to each other in this way are the same distance away from zero.

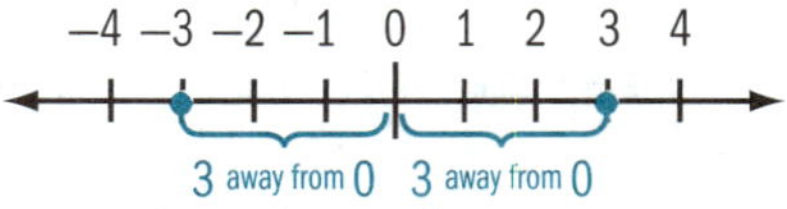


Zero is neither a positive number nor a negative number. It corresponds to itself.

In this book, we'll write the minus sign in front of negative numbers at the middle of the line, like we've been doing. But in some books, negative numbers are written like this: $^{-}4$ and $^{-}29$. The minus sign is raised and smaller in order to distinguish it from the regular subtraction sign. In these cases, it's called a **negative sign**.

For more on negative numbers, see Chapter 14. Until then, we'll be working with positive numbers only.

## INTEGERS

The **integers** are all the natural numbers, their negatives, and zero. All the numbers we've talked about so far are integers.

Numbers such as $\frac{1}{2}$, which are written as fractions when reduced to lowest terms, are not integers (see Chapter 7). Neither are numbers such as 4.78 or 0.3, which have a decimal part (see Chapter 10).



# DIGITS

A **digit** is just a symbol that helps represent numbers.

Our number system uses only ten **digits** to express any number. Here are all of them:

$$0, 1, 2, 3, 4, 5, 6, 7, 8, 9$$

That's it!

Any number can be written using these ten digits—small numbers such as 0 or 1, large numbers such as 234,676,900, fractions such as $\frac{3}{4}$, decimals such as 1.34, and many, many others.

## DIGITS VS. NUMBERS

A digit is a symbol that we use to write numbers down. A **number**, however, is the quantity that those digits represent.

The difference between digits and numbers is similar to the difference between letters and words. The word *a*, as in "Kyle Thomas is *a* really cute guy," is written with the letter *a*. The word *guy* uses the letters *g*, *u*, and *y*. Similarly, the number 13 is written using the digits 1 and 3, and the number 8 is written using the digit 8.

There are only ten digits, but infinitely many numbers. How can such a multitude of numbers all be written with the same ten digits? Well, to continue the analogy, any English word can be written down using only twenty-six letters. And the order of the letters matters very much. The words *no* and *on* are written using the same two letters—but they mean different things because the letters are arranged differently.

Similarly, the order of the digits in a number matters very much. The numbers 17 and 71 are written using the same two digits but mean very different things. I'm seventeen years old now, and it's so hard to believe that I will really be seventy-one someday. Ugh.

# PLACE VALUE

The location of a digit within a particular number determines how much that digit is worth. Each location—last, second to last, third to last—has a **place value**.

Take a look at the digits of the number 1467:

| 1 | 4 | 6 | 7 |
|---|---|---|---|
| thousands' place | hundreds' place | tens' place | ones' place |

| | |
|---|---|
| **Ones' place:** | The digit $7$ is in the ones' place.<br>The place value of the last digit is $1$. |
| **Tens' place:** | The digit $6$ is in the tens' place.<br>The place value of the second-to-last digit is $10$. |
| **Hundreds' place:** | The digit $4$ is in the hundreds' place.<br>The place value of the third-to-last digit is $100$. |
| **Thousands' place:** | The digit $1$ is in the thousands' place.<br>The place value of the fourth-to-last digit is $1000$. |

## SAME DIGIT, DIFFERENT PLACE VALUE

The numbers $311$, $131$, and $113$ are all written using one $3$ and two $1$s. But the value represented is very different. Let's look at the role of the digit $3$ in the three numbers.

In the number $311$, the digit $3$ is worth $3 \times 100$, or $300$, because it is in the third-to-last position. The place value of the third-to-last digit is $100$. This slot is called the **hundreds' place**.

In the number $131$, the digit $3$ is worth $3 \times 10$, or $30$, because it is in the second-to-last position. The place value of the second-to-last digit is $10$. This slot is called the **tens' place**.

In $113$, the digit $3$ is worth only $3 \times 1$, or $3$, because it is the last digit in the number. The place value of the last digit is $1$. This last slot is called the **ones' place**, or sometimes, the **units' place**.

## READING AND WRITING LARGE NUMBERS

In numbers longer than four digits, commas separate off each group of three digits, starting from the right. These groups are read off together. Here is the number 4,803,170,562 with its place values.

The number is read as "four billion, eight hundred and three million, one hundred and seventy thousand, five hundred and sixty-two."

# ROUNDING

Sometimes you don't need an *exact* number. For example, if Billy Idol had sold 3,008,432 copies of his hit album *Rebel Yell* in 1985, he would have said, "Eh, bloke, I sold three million copies of *Rebel Yell*! Triple platinum! Let's pick up some strippers and get blind!"

Billy rounded to the nearest million: 3,008,432 rounds down to 3,000,000.

## ROUND NUMBERS

A **round number** ends with one or several zeros. **To round** a number is to approximate it with the closest round number. The nearby round number that you pick depends on the context. Billy could have rounded 3,008,432 down to 3,008,000 instead of to 3,000,000, but "Eh, bloke, I just sold three million, eight thousand copies of *Rebel Yell*! That's triple platinum plus eight thousand! Let's pick up some strippers and get blind!" just doesn't have the same impact, so he chose not to be that specific.

## ROUNDING OFF: FIRST EXAMPLES

To round a number to, say, the nearest hundred . . .

- Find the digit in the hundreds' place.
- Look at the digit immediately to the right (the tens' digit).
- If the tens' digit is $0, 1, 2, 3$, or $4$, **round down**: keep the hundreds' digit and replace everything to the right with zeros. If the tens' digit is $5, 6, 7, 8$, or $9$, **round up**: increase the hundreds' digit by $1$, and replace all the digits to the right with zeros.

For example, let's round $7342$ to the nearest hundreds' place:

The $3$ is in the hundreds' place, and immediately to its right is a $4$. Since $4$ is small, round down: $7342$ becomes $7300$.

Another example: what's 379 to the nearest hundred?

The digit to the right of the hundreds' digit is a $7$, so we round up and add $1$ to the $3$ in the hundreds' place. We get $400$.

If the hundreds' digit is a $9$ and you end up having to round up, adjust everything accordingly. For example, what is $3965$ to the nearest hundred?

Well, the hundreds' digit is the $9$; to the right is a $6$. So we round up: the $9$ in the hundreds' place becomes a $0$ and the $3$ in the thousands' place becomes a $4$. So $3965$ to the nearest hundred is $4000$.

## ROUNDING OFF: MORE EXAMPLES

In general, to round off, find the digit immediately to the right of the place you're rounding to. If the digit is $4$ or smaller, round down; if it's $5$ or more, round up.

Let's round 379 to the nearest tens' place:

The tens' digit is the $7$; to the right is a $9$. Round up to get $380$.

Round 3965 to the nearest thousands' place:

The thousands' digit is the $3$; to the right is a $9$. Round up to get $4000$. If you remember, we got the same answer when we rounded $3965$ to the nearest hundreds' place. That's okay: it just means that $3965$ is pretty close to $4000$.

### ROUNDING DOWN AND ROUNDING UP

Sometimes we want to approximate a number with the closest *smaller* round number. This is called **rounding down**. We round down when we want to make sure not to overestimate.

To round down, just zero off every digit to the right of the place you're rounding down to. For example, 3489 rounded down to the nearest tens' place is 3480. The same number rounded down to the nearest hundred is 3400.

On the other hand, sometimes we want to approximate a number with the closest larger round number. This is called **rounding up**. We round up when we want to make sure not to underestimate.

To round up, increment the digit in the place you're rounding up to by 1; replace all the digits to the right by zeros. So 3489 rounded up to the nearest tens' place is 3490. The same number rounded up to the nearest hundred is 3500.

## INEQUALITY SIGNS

Just one more thing and we're done with all this orientation stuff. Let's race right through it. When writing math, people sometimes use various funny symbols just to save time. I completely agree: it's funny and weird. But so it goes.

Anyway, here are the symbols for comparing two numbers.

### EQUAL TO: =

The values on either side of the symbol are equal.

$1 = 1$ $455 = 455$ $1 + 1 = 2$ $14 = 10 + 4$

### LESS THAN: <

The value before the symbol is less than the value after.

$1 < 2$ $0 < 4$ $2 < 4{,}748$ $5 < 2 + 4$ $3 + 5 < 10$

### GREATER THAN: >

The value before the symbol is greater than the value after.

$2 > 0$ $49 > 2$ $3 + 5 > 4$ $3 > 1 + 1$

## LESS THAN OR EQUAL TO: ≤

The value before the symbol is not more than the value after.

$2 \leq 2$ $2 \leq 3$ $2 \leq 1 + 1$ $2 \leq 1 + 3$ $1 + 3 \leq 4$

## GREATER THAN OR EQUAL TO: ≥

The value before the symbol is not less than the value after.

$4 \geq 3$ $3 \geq 3$ $4 + 3 \geq 6$ $9 \geq 2 + 7$

## ME SO HUNGRY

If you get confused which way the sign points, keep in mind that the smaller, pointy end points toward the smaller number.

Also, if you think of the inequality sign as the mouth of a little creature, the mouth wants to eat the larger number. Like so:

# YOUR TURN

Solutions start on page 265.

1. Name the digit in

   (a) the ones' place of 1988 (That's the year I was born!)
   (b) the hundreds' place of 987,654
   (c) the ten-millions' place of 123,456,789,901

2. The number 0 is (pick all that apply)

   (a) a natural number
   (b) a whole number
   (c) an integer
   (d) a positive number
   (e) a negative number
   (f) a rational number
   (g) an irrational number

3. Round the number 134,990

   (a) to the nearest ten
   (b) to the nearest hundred
   (c) to the nearest thousand
   (d) to the nearest ten thousand
   (e) to the nearest hundred thousand

4. Read 123,456,789,901 out loud.

5. Use decimal digits to write the number three billion, one hundred and forty-one million, five hundred and ninety-two thousand, six hundred and fifty-three.

6. True or false?

   (a) $46 < 56$
   (b) $46 > 56 - 10$
   (c) $56 \geq 46$
   (d) $46 < 46$
   (e) $56 \leq 56$

# 2

# CHAPTER 2
# ADDITION AND SUBTRACTION



## HERE'S THE SCOOP

"Great, so now I understand *natural* numbers," Rebecca said. "But I still don't understand your *unnatural* attraction to Kyle Thomas."

"Clever wordplay with *unnatural* and *natural*, Rebecca!" I mean, who asked her? It's none of her business, right? So I have the hots for the new British punk rocker at school. Is that really so hard to fathom?

"I really think you should *rethink* Kyle Thomas," Rebecca continued.

"Hey, people told me to *rethink* you when you showed up in town last year," I countered. "I mean, what is up with your black clothes and that bag with a big BM monogram on it? Your name is Rebecca Romaine! RR!"

"You're right. I'm sorry. Let's just drop it," said Rebecca icily.

"Plus, you claim to have parents, but I've never seen them. And what's going on with your house? How come you never invite me over?"

"I said, let's drop it!"

"Okay! Touchy, touchy," I said. I mean, really, she is so weird sometimes. Verging on creepy-mysterious, if you ask me. But you didn't.

"Can't we just go back to studying math and stalking Kyle?" Rebecca asked. "Now that I have a better grasp on numbers, I'm really itching to use my newfound knowledge on tricky math thingies like adding and subtracting."

"Is that Kyle?" I blurted out. Rebecca whipped her head around.

"No," she said.

"Okay, adding and subtracting it is, then." I sighed. Where was he?

# ADDITION: THE BASICS

In **adding** two positive numbers, we combine them together to make a larger number, which we call their **sum**.

## ADDITION: TALKING AND WRITING

There are two common ways to write down addition. The first fits into one line:

$$1 + 3 = 4$$

We read this as "one plus three equals four." Or, more casually, "three added to one is four."

In the equation above, $1$ and $3$ are **addends**. (You'll rarely see this word, but it's nice to be able to recognize it.) The number $4$ is their **sum**. We also say that the expression $1 + 3$ is a sum. The symbol $+$ is called a **plus sign**.

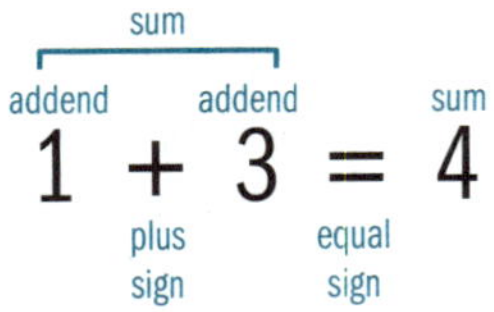


The other way of expressing addition is in a stack:

$$\begin{array}{r} 1 \\ +\,3 \\ \hline 4 \end{array}$$

The terminology is the same: $1$ and $3$ are addends, and $4$ is their sum.

## ADDING ON THE NUMBER LINE

On the number line, adding a positive number means moving to the right. Here's an example:

$$2 + 5 = 7$$

This means that we start at $2$ and move $5$ to the right, to end up at $7$. Like this:

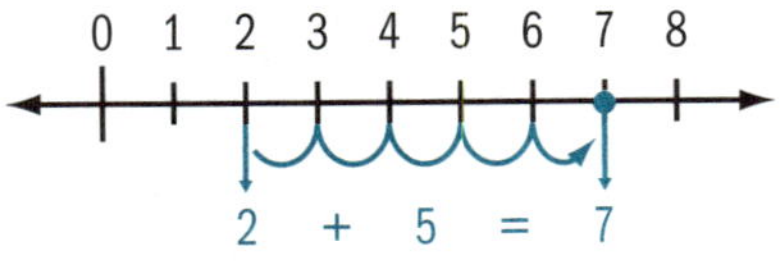

## ORDER DOESN'T MATTER!

An excellent thing about addition is that it doesn't matter which of the two numbers that we're adding comes first. So if we add $5 + 2$ instead of $2 + 5$, we get the same thing:

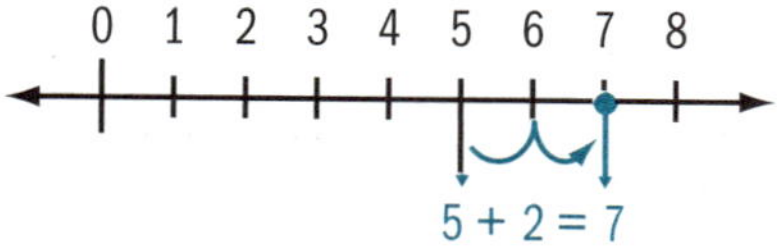


So $2 + 5 = 7$ and $5 + 2 = 7$. Convenient. We can say that $2 + 5 = 5 + 2$.

The fact the we can change the order of the addends is a nice property of addition. We say that addition is **commutative**. The $2$ and the $5$ can move past, or "commute" with, each other.

## ADDING IN THE REAL WORLD

Billy Idol is so adorable! I've been collecting *Rebel Yell* posters for years. I have $2$ posters on the wall with the window and $4$ more on the wall to the right. How many posters do I have altogether?

That's an addition problem: I have to add $2 + 4$. I can use the number line to figure it out:

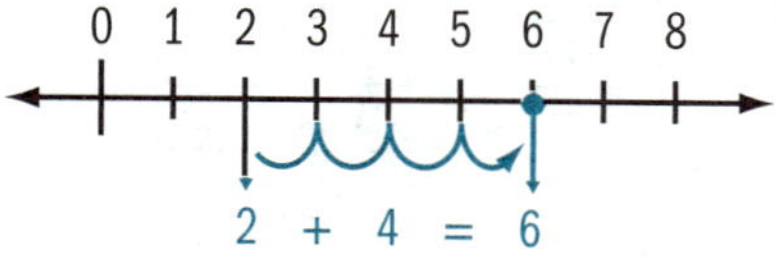


I have a total of $2 + 4 = 6$ *Rebel Yell* posters on my walls. I love Billy!

People add stuff all the time. Sometimes we add real things, like posters. But we can also add less concrete stuff—for example, money that someone owes you.

Rebecca is so peculiar: she always has tons of cash with her. I'm always broke, so she often has to spot me at the Finer Diner. Last week, I borrowed $\$3$, and then yesterday she loaned me another $\$6$. How much money do I owe her now?

Another addition problem! To figure it out, add $\$3 + \$6 = \$9$. So I owe her nine bucks. Crap.

We can even add more-abstract things. Yesterday, I passed Kyle in the hall six times before lunch (I know his schedule by heart, naturally), once during lunch, and then only twice after lunch. So how many times did I see Kyle yesterday?

This time we have to add *three* numbers. I passed him a total of $6 + 1 + 2 = 9$ times during the school day. Tragically, it's never enough.

## ADDING SMALL NUMBERS IN YOUR HEAD

Any addition problem with whole numbers can be done on the number line. But when the numbers are large, it would take a really, really long time. Imagine computing $34 + 27$. You'd have to start at $34$ and move over to the right $27$ times. Exhausting. Fortunately, there are various tricks for adding quickly, without drawing an enormous number line.

The first thing to do is a little tedious but crucial. You have to (gulp) memorize all the sums for small numbers: $1 + 1 = 2$, and $1 + 2 = 3$, and $1 + 4 = 5$, and $2 + 2 = 4$, and so on up through $9 + 9 = 18$. Unbelievably boring, I know. I'm sorry.

Luckily you probably already know most of these sums. But because they're so important for adding and subtracting, and multiplying, and everything else, I'm going to give you all of them here, arranged in a table.

| + | 0 | 1 | 2 | 3 | 4 | 5 | 6 | 7 | 8 | 9 |
|---|---|---|---|---|---|---|---|---|---|---|
| **0** | 0 | 1 | 2 | 3 | 4 | 5 | 6 | 7 | 8 | 9 |
| **1** | 1 | 2 | 3 | 4 | 5 | 6 | 7 | 8 | 9 | 10 |
| **2** | 2 | 3 | 4 | 5 | 6 | 7 | 8 | 9 | 10 | 11 |
| **3** | 3 | 4 | 5 | 6 | 7 | 8 | 9 | 10 | 11 | 12 |
| **4** | 4 | 5 | 6 | 7 | 8 | 9 | 10 | 11 | 12 | 13 |
| **5** | 5 | 6 | 7 | 8 | 9 | 10 | 11 | 12 | 13 | 14 |
| **6** | 6 | 7 | 8 | 9 | 10 | 11 | 12 | 13 | 14 | 15 |
| **7** | 7 | 8 | 9 | 10 | 11 | 12 | 13 | 14 | 15 | 16 |
| **8** | 8 | 9 | 10 | 11 | 12 | 13 | 14 | 15 | 16 | 17 |
| **9** | 9 | 10 | 11 | 12 | 13 | 14 | 15 | 16 | 17 | 18 |

Here's how the table works. To find out the sum of two numbers, pick the first number's row and the second number's column, and see where they intersect. So the answer to $6 + 7$ is in the square where row $6$ and column $7$ meet. According to the table, $6 + 7 = 13$.

| + | 0 | 1 | 2 | 3 | 4 | 5 | 6 | 7 | 8 | 9 |
|---|---|---|---|---|---|---|---|---|---|---|
| 0 | 0 | 1 | 2 | 3 | 4 | 5 | 6 | 7 | 8 | 9 |
| 1 | 1 | 2 | 3 | 4 | 5 | 6 | 7 | 8 | 9 | 10 |
| 2 | 2 | 3 | 4 | 5 | 6 | 7 | 8 | 9 | 10 | 11 |
| 3 | 3 | 4 | 5 | 6 | 7 | 8 | 9 | 10 | 11 | 12 |
| 4 | 4 | 5 | 6 | 7 | 8 | 9 | 10 | 11 | 12 | 13 |
| 5 | 5 | 6 | 7 | 8 | 9 | 10 | 11 | 12 | 13 | 14 |
| 6 | 6 | 7 | 8 | 9 | 10 | 11 | 12 | 13 | 14 | 15 |
| 7 | 7 | 8 | 9 | 10 | 11 | 12 | 13 | 14 | 15 | 16 |
| 8 | 8 | 9 | 10 | 11 | 12 | 13 | 14 | 15 | 16 | 17 |
| 9 | 9 | 10 | 11 | 12 | 13 | 14 | 15 | 16 | 17 | 18 |

The number 13 is also in the intersection of row 7 and column 6. That's because 7 + 6 is also equal to 13. The fact that the order of the addends doesn't matter translates into the fact that the addition table is symmetric on both sides of its diagonal.

# ADDING IN A STACK

Now that you know all the sums of small numbers, you can add larger numbers without too much trouble. The nicest way to keep track of everything is to stack the numbers up and add them digit by digit.

## STACK ADDITION: SIMPLEST EXAMPLE

Let's go through an example. What's 32 + 57?

First we stack the two numbers up, one on top of the other:

$$\begin{array}{r} 3\,2 \\ +\,5\,7 \\ \hline \end{array}$$

Now we're ready to start adding the numbers in each column. It's fastest and cleanest to add "backwards," from right to left.

**Add the ones' column:** 2 + 7 = 9. Write down 9 under the line.

$$\begin{array}{r} 3\,2 \\ +\,5\,7 \\ \hline 9 \end{array}$$

**Add the tens' column:** $3 + 5 = 8$. Write down $8$ under the line.

$$\begin{array}{r} 32 \\ +\ 57 \\ \hline 89 \end{array}$$

Finally, read off the answer under the line: $32 + 57 = 89$.

## STACK ADDITION: THREE-DIGIT EXAMPLE

Let's do another example, with larger numbers. What's $311 + 487$?

**Stack:**

$$\begin{array}{r} 311 \\ +\ 487 \\ \hline \end{array}$$

**Add the ones' column:** $1 + 7 = 8$.

$$\begin{array}{r} 311 \\ +\ 487 \\ \hline 8 \end{array}$$

**Add the tens' column:** $1 + 8 = 9$.

$$\begin{array}{r} 311 \\ +\ 487 \\ \hline 98 \end{array}$$

**Add the hundreds' column:** $3 + 4 = 7$.

$$\begin{array}{r} 311 \\ +\ 487 \\ \hline 798 \end{array}$$

Read off the answer: $311 + 487 = 798$.

## WATCH HOW YOU STACK!

I gave Rebecca another example: stack and add $201 + 94$.

Rebecca stacked like this:

$$\begin{array}{lll} & 2 & 0 & 1 \\ + & 9 & 4 & \\ \hline \end{array}$$

I stacked like this:

$$\begin{array}{r} 201 \\ +\ 94 \\ \hline \end{array}$$

Who is right? Trick question. I rule in math class, whereas Rebecca, bless her heart, is a bit of a math dummy. It's not that she's stupid, but she gets distracted so easily. She never pays attention in math class. She keeps scribbling furiously, though. Weird; I told you. And sometimes she just doesn't come to class at all.

So I'm right. But why? Because when you add each column, you want to be sure to add the ones to the ones, the tens to the tens, and so on. Otherwise you could add twenty to ten and get three hundred. Not good.

So you always have to make sure that all the digits line up. Like this:

| hundreds' digits | tens' digits | ones' digits |
|---|---|---|
| ↓ | ↓ | ↓ |
| 2 | 0 | 1 |
| + | 9 | 4 |



Let's just finish the problem. And now that you've got the stacking down, I'm going to stop with the vertical lines. They're a crutch. *Real* adders don't use 'em.

**Compute the ones' digit:** $1 + 4 = 5$.
**Compute the tens' digit:** $0 + 9 = 9$.

$$\begin{array}{r} 201 \\ +94 \\ \hline 95 \end{array}$$

**Compute the hundreds' digit:** There's only the $2$. So just copy it down.

$$\begin{array}{r} 201 \\ +94 \\ \hline 295 \end{array}$$

Done. We all rule!

## CARRYING AND REGROUPING

Deep breath. Time to talk about the trickiest thing in stack addition.

What's $57 + 26$?

**Stack:**

$$\begin{array}{r} 57 \\ +26 \\ \hline \end{array}$$

**Compute the ones' digit:** $7 + 6 = 13$.

Very true. But $13$ is not a digit. What to do?

Instead of writing down $13$, we write down $3$ in the ones' place and **carry** the $1$ to the next column over to the left. This means that we'll add $1$ to whatever we get for the tens' digit of the sum. To make sure that we remember to add $1$, we write a little "$1$" over the tens' column, like so:

$$\begin{array}{r} {\scriptstyle 1}\phantom{7} \\ 57 \\ +26 \\ \hline 3 \end{array}$$

Now, **compute the tens' digit:** $5 + 2 = 7$. Add the $1$ carried over from the ones' place to get $8$. Write it down.

$$\begin{array}{r} {\scriptstyle 1}\phantom{7} \\ 57 \\ +26 \\ \hline 83 \end{array}$$

Excellent! We get $57 + 26 = 83$.

All this carrying business is also called **regrouping**. That's because when we carry the $1$ in the problem above, we're essentially thinking of $13$ as $1 \,\vdots\, 3$, a $1$ in the tens' column and a $3$ in the ones' column. (That's just because $13$ is $10 + 3$.) So the $3$ is written down in the ones' place of the answer, and the $10$ is added to the tens—where it belongs. We're redistributing—"regrouping"—the sum of the ones' digits among the ones and the tens.

Let's go through another example: $958 + 76 = ?$

**Stack:**

$$\begin{array}{r} 958 \\ +76 \\ \hline \end{array}$$

**Compute the ones' digit:** $8 + 6 = 14$. Write down $4$; carry the $1$.

$$\begin{array}{r} {\scriptstyle 1\,} \\ 958 \\ +76 \\ \hline 4 \end{array}$$

**Compute the tens' digit:** $5 + 7 = 12$. Add $1$ to get $13$. We have to regroup *again*. Write down $3$; carry the $1$.

$$\begin{array}{r} {\scriptstyle 1\,1\,\,} \\ 958 \\ +76 \\ \hline 34 \end{array}$$

**Compute the hundreds' digit:** $1 + 9 = 10$. Nowhere left to carry; write down $10$.

$$\begin{array}{r} {\scriptstyle 1\,1\,\,} \\ 958 \\ +76 \\ \hline 1034 \end{array}$$

Done! The answer is $1034$.

The best part? That's all there is to know! You're ready to do addition problems with monstrously enormous numbers. There are some waiting for you at the end of the chapter.

## ADDING MORE THAN TWO NUMBERS

Adding three numbers is just like adding two numbers. You already know how to do this. But let's just do an example anyway.

$$\begin{array}{r} 972 \\ 371 \\ +2188 \\ \hline \end{array}$$

Notice how carefully I stacked: the hundreds' digits of the first two numbers–$9$ and $3$–line up with the $1$ in the hundreds' place of the third number.

**Compute the ones' digit:** $2 + 1 + 8 = 11$. Write down $1$; carry the $1$.

$$\begin{array}{r} {\scriptstyle 1}\phantom{0} \\ 972 \\ 371 \\ +2188 \\ \hline 1 \end{array}$$

**Compute the tens' digit:** $1 + 7 \ + 7 + 8 = 23$. Write down $3$; *carry the 2*. A bit unexpected, but can happen when adding three or more numbers.

$$\begin{array}{r} {\scriptstyle 2\ 1}\phantom{0} \\ 972 \\ 371 \\ +2188 \\ \hline 31 \end{array}$$

**Compute the hundreds' digit:** $2 + 9 + 3 + 1 = 15$. Write down $5$; carry the $1$.

$$\begin{array}{r} {\scriptstyle 1\ 2\ 1}\phantom{0} \\ 972 \\ 371 \\ +2188 \\ \hline 531 \end{array}$$

**Compute the thousands' digit**, including $1$ carried over from the hundreds: $1 + 2 = 3$.

$$\begin{array}{r} {\scriptstyle 1\ 2\ 1}\phantom{0} \\ 972 \\ 371 \\ +2188 \\ \hline 3531 \end{array}$$

Done! The answer is $3531$.

Easy enough, eh? The only thing to watch out for is the carry digit. When adding two numbers, the only digit you'd ever have to carry is $1$. When adding three numbers, you may have to carry a $1$ or a $2$. The more numbers you're adding, the larger the largest possible carry digit.

That's it for addition. You're a champ. Let's move on.

# SUBTRACTION

Subtraction and addition are closely related: you can think of subtracting as undoing addition. The great part for you? Not that much more to learn.

## SUBTRACTION: TALKING AND WRITING

Just like addition, subtraction can be written horizontally on one line or vertically in a stack. On one line, a subtraction equation looks like this:

$$9 - 3 = 6$$

We read this as "nine minus three equals six" or "nine take away three is six" or even "the difference of nine and three is six."

In the equation above, $9$ is the **minuend**, $3$ is the **subtrahend**, and $6$ is their **difference**. (Again, the words *minuend* and *subtrahend* are pretty obscure these days.) We can also call the expression $9 - 3$ a difference. The symbol – is called the **minus sign**.

difference (bracket over 9 – 3)

minuend · subtrahend · difference

$$9 - 3 = 6$$

minus sign · equal sign

Stack subtraction looks like this:

$$\begin{array}{r} 9 \\ -3 \\ \hline 6 \end{array}$$

If you're working with positive numbers, the minuend should always be greater than the subtrahend. In this case, 9 is greater than 3.

## SUBTRACTING ON THE NUMBER LINE

On the number line, subtracting a positive number means moving to the left. For example,

$$8 - 3 = 5$$

We start at $8$ and move $3$ to the left, to end up at $5$. Like this:

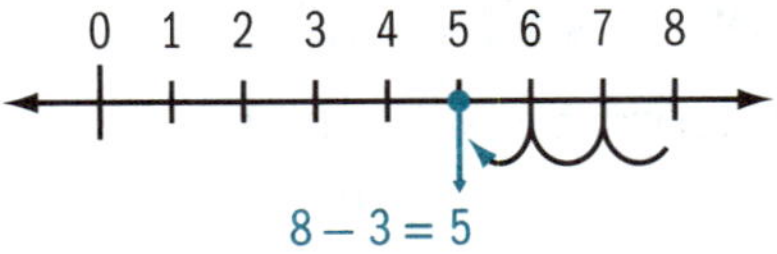


## THE ORDER MATTERS!

One big difference between addition and subtraction: in subtraction, the order matters very much. In fact, if you're sticking with positive numbers, you *can't* switch the order. You can see on the number line that computing $3 - 8$ is impossible—unless we veer to the left of zero. But that's for another chapter.

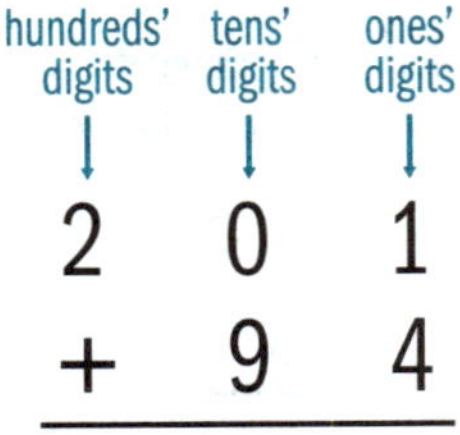


Why does this make sense? Positive numbers count real-life things, like Billy Idol posters or the number of times I run into Kyle. Or pencils. Say Rebecca has $8$ pencils. If she gives me $3$ of them, she has $5$ left. But if she only has $3$ pencils, she can't give me $8$ of them. She just doesn't have enough. When working with real-life things or positive numbers, the number you subtract from has to be at least as big as the number being subtracted.

## SUBTRACTING SMALL NUMBERS

So how do we actually subtract numbers?

If the numbers are small, we rely on what we already know about addition. To determine $15 - 7$, think of a number that, when added to $7$, gives you $15$. The number $8$ works because $7 + 8 = 15$. Great. So $15 - 7 = 8$. That's it.

Try again: $12 - 9$. This is the same as figuring out what number, when added to $9$, gives you $12$. If you forget, use the addition table on page 18. Since $9 + 3 = 12$, the number $3$ works. So

$$12 - 9 = 3.$$

## SUBTRACTING IN A STACK

When the numbers are larger, we subtract in a stack, digit by digit. Again, stack carefully—make sure that the ones' digits line up, and the rest will follow. Let's do an example.

$$\begin{array}{r} 249 \\ -32 \\ \hline \end{array}$$

**Compute the ones' digit:** $9 - 2 = 7$.

$$\begin{array}{r} 249 \\ -32 \\ \hline 7 \end{array}$$

**Compute the tens' digit:** $4 - 3 = 1$.

$$\begin{array}{r} 249 \\ -32 \\ \hline 17 \end{array}$$

**Compute the hundreds' digit:** Nothing to subtract, so just bring down the $2$.

$$\begin{array}{r} 249 \\ -32 \\ \hline 217 \end{array}$$

## BORROWING AND REGROUPING

Now comes the trickiest part of subtraction. Let's walk through another example.

$$\begin{array}{r} 63 \\ -27 \\ \hline \end{array}$$

**Compute the ones' digit:** $3 - 7 = \ldots$ Hmm. You can't subtract $7$ from $3$.

However, you *can* subtract $7$ from $13$. So we **borrow** a ten from the tens' digit. The $6$ in the tens' place becomes a $5$, and the $3$ in the ones' place becomes a $13$.

To keep track of all these changes, we usually indicate them on the problem: cross out the $6$ and replace with a small $5$; cross out the $3$ and replace with a small $13$. Like this:

$$\begin{array}{r} {\scriptstyle 5\;13} \\ \not{6}\not{3} \\ -27 \\ \hline \end{array}$$

Now compute the ones' digit: $13 - 7 = 6$.

$$\begin{array}{r} {\scriptstyle 5\;13} \\ \not{6}\not{3} \\ -27 \\ \hline 6 \end{array}$$

To finish up, **compute the tens' digit**, remembering that you've borrowed: $5 - 2 = 3$.

$$\begin{array}{r} {\scriptstyle 5\;13} \\ \not{6}\not{3} \\ -27 \\ \hline 36 \end{array}$$

Awesome: $63 - 27 = 36$.

Like carrying, borrowing is sometimes also called **regrouping**. That's because we're reconsidering the way we think about the minuend. Normally, we think of $63$ as $60 + 3$ and perform the subtraction separately on each digit. But when the ones' digit of the subtrahend is too large, we have to "regroup" the digits and think of $63$ as $50 + 13$. Then we can find the difference, digit by digit.

## EXHAUSTING EXAMPLE

Let's do a more complicated example:

$$\begin{array}{r} 23170 \\ -19074 \\ \hline \end{array}$$

Daunting, huh? We just have to keep chugging. Watch out: we'll have to borrow more than once.

**Compute the ones' digit:** $0 - 4$ is impossible, so borrow a $1$ from the tens' place. The $7$ in the tens' place becomes a $6$; the $0$ in the ones' place becomes a $10$. Finally, $10 - 4 = 6$.

$$\begin{array}{r} {\scriptstyle 6\ 10} \\ 231\cancel{7}\cancel{0} \\ -19074 \\ \hline 6 \end{array}$$

**Compute the tens' digit**, remembering that you've borrowed: $6 - 7$ is impossible. Borrow a $1$ from the hundreds' place. The $1$ in the hundreds' place becomes a $0$; the $6$ in the tens' place becomes a $16$. And the subtraction gives you $16 - 7 = 9$. Whew.

$$\begin{array}{r} {\scriptstyle 16\ \ } \\ {\scriptstyle 0\ \cancel{6}\ 10} \\ 23\cancel{1}\cancel{7}\cancel{0} \\ -19074 \\ \hline 96 \end{array}$$

It's a little annoying that we've now had to change the tens' digit twice, but so it goes.

**Compute the hundreds' digit:** $0 - 0 = 0$. We lucked out there.

$$\begin{array}{r} {\scriptstyle 16\ \ } \\ {\scriptstyle 0\ \cancel{6}\ 10} \\ 23\cancel{1}\cancel{7}\cancel{0} \\ -19074 \\ \hline 096 \end{array}$$

**Compute the thousands' digit:** $3 - 9$ is impossible. Uh-oh. We have to borrow again. The $2$ in the ten-thousands' place becomes a $1$; the $3$ in the thousands' place becomes a $13$. The subtraction gives us $13 - 9 = 4$.

$$\begin{array}{r} 16 \\ 1\ 13\ 0\ \not{6}\ 10 \\ \not{2}\not{3}\not{1}\not{7}\not{0} \\ -19074 \\ \hline 4096 \end{array}$$



Finally, **compute the ten-thousands' digit:** $1 - 1 = 0$. There are no more digits left, so you don't have to write anything down. The answer is $4096$.

That was a real doozy! And pretty yucky to boot. But there was nothing new there; you just have to crank it out. Keep moving right to left, and keep borrowing if you must.

## BORROWING FROM A ZERO

But what happens if you run into a zero on your borrowing spree? Last example, I promise.

$$\begin{array}{r} 402 \\ -75 \\ \hline \end{array}$$

**Compute the ones' digit:** Can't do $2 - 5$, so we want to borrow from the tens' place—but there's a big fat $0$ sitting there. So we borrow from the hundreds instead.

Here's how to do it: borrow $1$ from the $4$ in the hundreds' place so the $4$ becomes a $3$, change the $0$ in the tens' place to a $9$, and give a generous ten to the ones' place, making the $2$ into a $12$:

$$\begin{array}{r} 3\ 9\ 12 \\ \not{4}\not{0}\not{2} \\ -75 \\ \hline \end{array}$$

Now you can do the subtraction, digit by digit: in the ones' place, $12 - 5 = 7$. **Compute the tens' digit:** $9 - 7 = 2$.

**Compute the hundreds' digit:** Nothing to subtract, so just bring down the 3.

$$\begin{array}{r} {\scriptstyle 3\ 9\ 12} \\ \not{4}\not{0}\not{2} \\ -75 \\ \hline 327 \end{array}$$

Curious why this works? Effectively, we're borrowing from the hundreds' place to enrich the tens' place—making the 0 in the tens' place into a 10—and *then* borrowing from the new 10 in the tens' place, which makes the 10 into a 9. The whole thing can also be written like this:

$$\begin{array}{r} {\scriptstyle 9\quad\ } \\ {\scriptstyle 3\ \not{10}\ 12} \\ \not{4}\not{0}\not{2} \\ -75 \\ \hline 327 \end{array}$$

But that's pretty messy, so we usually skip the intermediate step.

## CHECKING YOUR WORK

One last thing: because subtracting is the same thing as undoing addition, you can check the answer to a subtraction problem by doing some addition.

To check the answer to any subtraction problem, add number you subtracted to the difference. You should get the number you subtracted *from*.

So to check that $402 - 75 = 327$, we compute $327 + 75$. We'd better get 402, or else something's wrong.

$$\begin{array}{r} {\scriptstyle 1\ 1\ \ } \\ 327 \\ +75 \\ \hline 402 \end{array}$$

Lucky us! It seems to have worked out. If we hadn't gotten 402, we would have known that we made a mistake somewhere.

Similarly, you can check addition problems with subtraction. Subtract either one of the addends from the sum; you should get the other addend.

# YOUR TURN

Solutions start on page 265.



1. In $98 + 22 = 120$, which expression represents the sum? Pick all that apply.

   (a) $98$
   (b) $22$
   (c) $120$
   (d) $+22$
   (e) $98 + 22$
   (f) $22 = 120$

2. Find the sum: $131 + 7245 =$

3. What is the sum when $3947$ is added to $9273$?

4. Find the difference: $927 - 82 =$

5. What is the difference when $127$ is subtracted from $301$?

6. Kyle has a gift certificate to Barnes & Noble for \$50. He uses it to pay \$34 for a three-volume book on the world's greatest lead guitarists. How much money is remaining on the gift card after the purchase?

7. Maplewood High had $85$ tenured teachers last year. Over the summer, $8$ teachers lost their tenure and $3$ teachers got tenure. How many tenured teachers are there at Maplewood High this year?

8. This year at Maplewood High there are $295$ freshmen, $310$ sophomores, $260$ juniors, $232$ seniors, and $6$ special students. How many students are enrolled in our high school this year?

9. Find the mystery number.

$$\begin{array}{r} 384 \\ +??? \\ \hline 900 \end{array}$$

# 3

# CHAPTER 3
## MULTIPLICATION



### KYLE SAYS "SHITE"

If he didn't walk by the Finer Diner soon, Kyle was going to be late for math class. Rebecca was getting antsy.

"I have to go to class *now*, Shanon. Unlike you, I'm actually taking it to learn *math*."

"Fine, go! I'm just going to wait here a few more minutes. Kyle has to walk by. Class starts really soon!"

"Yes, it does. See you there."

"Fine, bye!"

Rebecca Romaine took her big black BM bag and left me all alone. All the better for me! This way Kyle won't be tempted to talk to Rebecca when I ambush him. Rebecca is always quick to come up with reasons why I'm overdoing my pursuit of Kyle. She's always like, "He's not so great," and, "Billy Idol sucks and so does Johnny London." Which are both obviously not true!

I think she's just trying to throw me off her scent! I'm worried that she really likes Kyle, and if she does, I'm in big trouble, because she's the cutest girl in school and I am, well, not.

Omigod. There he is!

"Hi, Kyle!"

"Um, hi?"

"It's me, Shanon, from math. Are you on your way there?"

"Um, yeah?"

"Great! Me too. Hold on, I just have to pay for this cinnamon bun and grapefruit juice."

"Cinnamon bun and grapefruit juice, mate? That's an odd meal."

"Oh, hah, think so? Well, it's not mine, it's Rebecca Romaine's. She was just here and she ordered it, and she's a freak, eating cinnamon buns and, well, grapefruit juice, although you don't really eat grapefruit juice, now do you, you drink it, really. . . ."

"Okay, whatever you say, mate."

"There. Paid. I always leave a big fat tip. See—three bucks! . . . So . . . how are you doing in class?"

"Well, I understand adding and subtracting, but these multiplication tables are *shite*!"

"Shite! Right! I love how you British people make bad words sound even cooler."

"Are you good at multiplying, mate?"

"Are you coming on to me, Kyle?"

"Um, no? We're talking about math, right?"

"Oh yes, *multiplication*! Yeah, I'm okay at it. I've done some multiplying in my day. It's sort of like adding, only better. Um . . . Let me explain as we walk, side by side, down the sidewalk of life."

# MULTIPLICATION: THE BASICS

**Multiplying** a number by another number is the same thing as adding the first number to itself several times. Not hard.

So $3 \times 4$ means "3 added to itself 4 times," or $3 + 3 + 3 + 3$:

$$3 \times 4 = \overbrace{3 + 3 + 3 + 3}^{\text{4 times}} = 12$$

## WRITING IT DOWN

Multiplication can be expressed in a number of ways. And people often get pretty careless about writing it down. It's a bit annoying, but you can usually figure out what's going on.

**Cross:** $3 \times 4 = 12$

This is the notation we'll use most of the time.

**Dot:** $3 \cdot 4 = 12$

**Parentheses:** $3(4) = 12$

**Two parenthesis pairs:** $(3)(4) = 12$

**Stack:**
$$\begin{array}{r} 3 \\ \times 4 \\ \hline 12 \end{array}$$

## TALKING ABOUT IT

People also say all sorts of things when they're talking about multiplication.

So for $5 \times 2 = 10$, you can say "five times two equals ten" or "five multiplied by two makes ten" or even "twice five is ten." Either way, you're adding two fives and getting ten. Easy enough.

In the equation $5 \times 2 = 10$, the number $5$ is the **multiplicand**, $2$ is the **multiplier**, and $10$ is the **product**. You can forget the first two; the word *product* is the only one that you need to worry about.

## ORDER DOESN'T MATTER

It turns out that the order of the two numbers being multiplied together doesn't matter at all. The product is the same. So $3 \times 4$ is the same as $4 \times 3$. Both are equal to $12$.

This is truly a gift. It makes all sorts of mathy things much, much easier.

Why is this true? It's not completely obvious: $3 \times 4$ is adding four $3$s together, whereas $4 \times 3$ is adding three $4$s together. So why should $3 + 3 + 3 + 3$ be the same thing as $4 + 4 + 4$? Take a look at this arrangement of valentines.

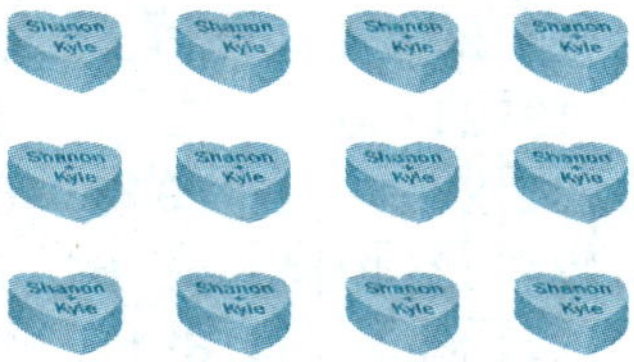

How many are there? (Well, it's obvious that there are twelve of them, but stay with me.)

On one hand, there are four columns with three valentines in each. So there's a total of $3 + 3 + 3 + 3 = 3 \times 4$ valentines. On the other hand, there are three rows of four valentines each. That makes a total of $4 + 4 + 4 = 4 \times 3$ valentines. So $3 \times 4$ ends up being the same as $4 \times 3$, which is extremely convenient in math.

Incidentally, this neat trick with switching the order of the numbers being multiplied has a name: we say that multiplication is **commutative**. (Addition is commutative, too, remember?) We use that word because the numbers can move past, or "commute" with, each other.

# TIMES TABLES

We've gone over everything you need to know about multiplication—except how people actually do it. It's all well and good to add four 2s if you're doing a small problem like $2 \times 4$, but by the time we get to $7 \times 9$ or to $256 \times 1024$, it gets very tedious.

Most people grin and bear it: they memorize all the products of small numbers, from $1 \times 1 = 1$ to $9 \times 9 = 81$. All the small products are usually stored in a table, like the one below.

To survive in the math wilderness, you need to memorize the products of only the single-digit numbers, but I'm including 10 (because it's so easy) as well as 11 and 12 (because some teachers make students learn those too).

| × | **0** | **1** | **2** | **3** | **4** | **5** | **6** | **7** | **8** | **9** | **10** | **11** | **12** |
|---|---|---|---|---|---|---|---|---|---|---|---|---|---|
| **0** | 0 | 0 | 0 | 0 | 0 | 0 | 0 | 0 | 0 | 0 | 0 | 0 | 0 |
| **1** | 0 | 1 | 2 | 3 | 4 | 5 | 6 | 7 | 8 | 9 | 10 | 11 | 12 |
| **2** | 0 | 2 | 4 | 6 | 8 | 10 | 12 | 14 | 16 | 18 | 20 | 22 | 24 |
| **3** | 0 | 3 | 6 | 9 | 12 | 15 | 18 | 21 | 24 | 27 | 30 | 33 | 36 |
| **4** | 0 | 4 | 8 | 12 | 16 | 20 | 24 | 28 | 32 | 36 | 40 | 44 | 48 |
| **5** | 0 | 5 | 10 | 15 | 20 | 25 | 30 | 35 | 40 | 45 | 50 | 55 | 60 |
| **6** | 0 | 6 | 12 | 18 | 24 | 30 | 36 | 42 | 48 | 54 | 60 | 66 | 72 |
| **7** | 0 | 7 | 14 | 21 | 28 | 35 | 42 | 49 | 56 | 63 | 70 | 77 | 84 |
| **8** | 0 | 8 | 16 | 24 | 32 | 40 | 48 | 56 | 64 | 72 | 80 | 88 | 96 |
| **9** | 0 | 9 | 18 | 27 | 36 | 45 | 54 | 63 | 72 | 81 | 90 | 99 | 108 |
| **10** | 0 | 10 | 20 | 30 | 40 | 50 | 60 | 70 | 80 | 90 | 100 | 110 | 120 |
| **11** | 0 | 11 | 22 | 33 | 44 | 55 | 66 | 77 | 88 | 99 | 110 | 121 | 132 |
| **12** | 0 | 12 | 24 | 36 | 48 | 60 | 72 | 84 | 96 | 108 | 120 | 132 | 144 |

Learning all these products is nothing if not mind-bogglingly dull, but it really, really helps. Honestly, I recommend it. Otherwise you might get lost later.

## USING THE TABLE

The multiplication table works just like the addition table. To find out the product of two numbers, pick the first number's row and the second number's column, and see where they intersect. The answer to $6 \times 7$ is in the square where row 6 and column 7 meet. So $6 \times 7 = 42$. Take a look:

| × | 0 | 1 | 2 | 3 | 4 | 5 | 6 | 7 | 8 | 9 | 10 | 11 | 12 |
|---|---|---|---|---|---|---|---|---|---|---|---|---|---|
| 0 | 0 | 0 | 0 | 0 | 0 | 0 | 0 | 0 | 0 | 0 | 0 | 0 | 0 |
| 1 | 0 | 1 | 2 | 3 | 4 | 5 | 6 | 7 | 8 | 9 | 10 | 11 | 12 |
| 2 | 0 | 2 | 4 | 6 | 8 | 10 | 12 | 14 | 16 | 18 | 20 | 22 | 24 |
| 3 | 0 | 3 | 6 | 9 | 12 | 15 | 18 | 21 | 24 | 27 | 30 | 33 | 36 |
| 4 | 0 | 4 | 8 | 12 | 16 | 20 | 24 | 28 | 32 | 36 | 40 | 44 | 48 |
| 5 | 0 | 5 | 10 | 15 | 20 | 25 | 30 | 35 | 40 | 45 | 50 | 55 | 60 |
| 6 | 0 | 6 | 12 | 18 | 24 | 30 | 36 | 42 | 48 | 54 | 60 | 66 | 72 |
| 7 | 0 | 7 | 14 | 21 | 28 | 35 | 42 | 49 | 56 | 63 | 70 | 77 | 84 |
| 8 | 0 | 8 | 16 | 24 | 32 | 40 | 48 | 56 | 64 | 72 | 80 | 88 | 96 |
| 9 | 0 | 9 | 18 | 27 | 36 | 45 | 54 | 63 | 72 | 81 | 90 | 99 | 108 |
| 10 | 0 | 10 | 20 | 30 | 40 | 50 | 60 | 70 | 80 | 90 | 100 | 110 | 120 |
| 11 | 0 | 11 | 22 | 33 | 44 | 55 | 66 | 77 | 88 | 99 | 110 | 121 | 132 |
| 12 | 0 | 12 | 24 | 36 | 48 | 60 | 72 | 84 | 96 | 108 | 120 | 132 | 144 |

The number 42 is also in the intersection of row 7 and column 6. That's because $7 \times 6$ is also equal to 42. The order doesn't matter, remember? Multiplication is commutative.

In fact, because the order of multiplication doesn't matter, the two halves of the table split by the main diagonal are the same. If you fold the table along the diagonal, the two halves would match up. This table has *symmetry*.

## MULTIPLICATION TRICKS

I've got a few tricks to help you with memorizing the multiplication table. Some of these correspond to patterns that we can see in the multiplication table, which is a nice feature

- **Multiplying by 0:** Any number times 0 is 0. And 0 times any number is 0. This corresponds to the fact that all the numbers in the first column and in the first row of the table are all zeros.

  For example, $0 \times 3 = 0 + 0 + 0 = 0$. The other way is a little more confusing: $5 \times 0$ is adding zero 5s together. Since we're adding no 5s, we get zero: $5 \times 0 = 0$.

- **Multiplying by 1:** Any number times 1 is itself. And 1 times any number is that number again. This corresponds to the fact that all column 1 just repeats all the numbers in order. Row 1 does too.

  So $1 \times 3 = 1 + 1 + 1 = 3$. And $8 \times 1$ is adding up one 8, which gives you 8, like this: $8 \times 1 = 8$.
- **Multiplying by 10:** This one is really satisfying. To multiply any number by 10, simply write it down and then add on a 0 at the end. So $3 \times 10 = 30$ and $45 \times 10 = 450$. This works when multiplying 10 by any number, too: $10 \times 12 = 120$.
- **Multiplying by 9:** This one's my favorite and the only real trick here. To multiply any number by 9, look at your ten fingers, palms up. Bend down the finger that corresponds to the number that you want to multiply, counting from the left, and read off the answer, in two digits!

  For example, to compute $4 \times 9$, bend down the 4th finger (left-hand ring finger). There are 3 fingers to the left of the bend and 6 fingers to the right. The answer is 36!

  Nifty, eh? Try it yourself.

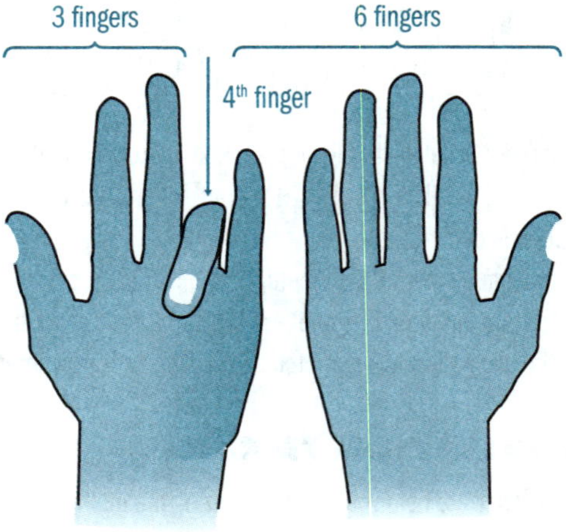


MULTIPLICATION

- **Multiplying by 5:** A little complicated, but it works nicely if you're a decent divider. (We'll go over division in the next chapter, so don't worry if you're not.) If a number is evenly divisible by $2$, then instead of multiplying by $5$, you can divide it by $2$ and add on a $0$ digit at the end.

  For example, let's try $6 \times 5$. Half of $6$ is $3$, and we attach a zero to make $30$. So $6 \times 5 = 30$. This works for larger numbers too. For example, $14 \times 5 = 70$: half of $14$ is $7$; add on a $0$ to make $70$.

# MULTIPLYING LARGER NUMBERS

Now that you're an ace at multiplying smaller numbers, let's move on to some real techniques. Larger numbers are multiplied in a stack, one digit at a time. As with addition, we write down the last digit of each product and carry the rest.

## MULTIPLYING BY A ONE-DIGIT NUMBER

Enough talk. Let's dig right in. We'll start with a really easy example just to get going. What's $231 \times 3$?

**Stack:** Proper stacking matters a lot less here than with addition, but it helps to do things neatly.

$$\begin{array}{r} 231 \\ \times 3 \\ \hline \end{array}$$

Now we work through the top number, right to left digit by digit, multiplying each one by the bottom number.

**Multiply the ones' digit:** $1 \times 3 = 3$. Write down $3$ in the ones' column under the line.

$$\begin{array}{r} 231 \\ \times 3 \\ \hline 3 \end{array}$$

**Multiply the tens' digit:** $3 \times 3 = 9$. Write it down in the tens' column.

$$\begin{array}{r} 231 \\ \times 3 \\ \hline 93 \end{array}$$

MULTIPLICATION

**Multiply the hundreds' digit:** $2 \times 3 = 6$. Write it down in the hundreds' column.

$$\begin{array}{r} 231 \\ \times\, 3 \\ \hline 693 \end{array}$$

Read off the answer: $231 \times 3 = 693$.

That was a piece of cake, right? Let's spice things up.

## MULTIPLYING WITH CARRYING

In the previous example, every product was a one-digit number, so we didn't have to carry. This time, get ready for some regrouping:

$$\begin{array}{r} 3065 \\ \times\quad 7 \\ \hline \end{array}$$

**Compute the ones' digit:** $5 \times 7 = 35$. Write down $5$; carry the $3$. To keep track of the digit you carry, write it down above the next column.

$$\begin{array}{r} {\scriptstyle 3}\; \\ 3065 \\ \times\quad 7 \\ \hline 5 \end{array}$$

**Compute the tens' digit** by first multiplying, then adding the carried-over digit: $6 \times 7 = 42$. Add $3$ to get $45$. Write down $5$, carry $4$.

$$\begin{array}{r} {\scriptstyle 4\,3}\;\; \\ 3065 \\ \times\quad 7 \\ \hline 55 \end{array}$$

**Compute the hundreds' digit** (multiply, add, write down): $0 \times 7 = 0$. Add $4$ to get $4$. Write down $4$. Nothing to carry.

$$\begin{array}{r} {\scriptstyle 4\,3}\;\; \\ 3065 \\ \times\quad 7 \\ \hline 455 \end{array}$$

**Compute the thousands' digit:** $3 \times 7 = 21$. Nothing to add on. No place to carry, so write down $21$ and read off the answer.

$$\begin{array}{r} \scriptstyle 43\phantom{0} \\ 3065 \\ \times\ \ 7 \\ \hline 21455 \end{array}$$

So $3065 \times 7 = 21{,}455$. That wasn't so bad.

## MULTIPLYING BY A TWO-DIGIT NUMBER

Do it digit by digit. First, multiply the whole first number by the ones' digit of the second number. Then multiply the whole first number by the tens' digit of the second number. Finally, add the results.

Let's do an example:

$$\begin{array}{r} 169 \\ \times 43 \\ \hline \end{array}$$

First, multiply $169$ by $3$ as you normally would. Ignore the $4$ in $43$ entirely. I'm going to go quickly to get to the important new stuff.

**169 × 3: ones' digit:** $9 \times 3 = 27$. Write $7$; carry $2$.

**169 × 3: tens' digit:** $6 \times 3 + 2 = 20$. Write $0$; carry $2$.

**169 × 3: hundreds' digit:** $1 \times 3 + 2 = 5$. Write $5$. This piece is done:

$$\begin{array}{r} \scriptstyle 22\phantom{0} \\ 169 \\ \times 43 \\ \hline 507 \end{array}$$

Now multiply $169$ by $4$, ignoring the $3$ this time. The only tricky part is that because the $4$ is in the tens' place (and so actually represents $40$), start writing the product in the tens' column, not the ones' column.

**169 × 4: *tens'* digit:** $9 \times 4 = 36$. Write $6$, carry $3$. To make sure you don't get confused with the old carry numbers, erase them or cross them out.

$$\begin{array}{r} \scriptstyle 3\phantom{00} \\ \scriptstyle \cancel{2}\cancel{2}\phantom{0} \\ 169 \\ \times 43 \\ \hline 507 \\ 6\phantom{0} \end{array}$$

**169 × 4: *hundreds'* digit:** $6 \times 4 + 3 = 27$. Write 7, carry 2.

$$\begin{array}{r} {\scriptstyle 2\,3} \\ {\scriptstyle \not{2}\,\not{2}} \\ 169 \\ \times\ 43 \\ \hline 507 \\ 76\phantom{0} \end{array}$$

**169 × 4: *thousands'* digit:** $1 \times 4 + 2 = 6$. Write down 6. Nothing to carry.

$$\begin{array}{r} {\scriptstyle 2\,3} \\ {\scriptstyle \not{2}\,\not{2}} \\ 169 \\ \times\ 43 \\ \hline 507 \\ 676\phantom{0} \end{array}$$

You're getting there. Last bit: add up the two partial-product rows to get the final answer. Be careful with how the digits are stacked here.

$$\begin{array}{r} 169 \\ \times\ 43 \\ \hline 507 \\ +676\phantom{0} \\ \hline 7267 \end{array}$$

That's it: $169 \times 43 = 7267$.

That was a little painful in execution, but the theory isn't that bad. Each digit of the second number gets its own row. Multiply out all the partial products, then add them up. That's all there is to know.

## MULTIPLYING BY A LARGE NUMBER

Like I said, you know all there is to know. But let's just do one more nasty example, just for kicks. (To keep things clean, I'm going to get rid of all the carry numbers after each row.)

$$\begin{array}{r} 3804 \\ \times\ 719 \\ \hline \end{array}$$

**Multiply by 9:**

$4 \times 9 = 36$. Write 6; carry 3.

$0 \times 9 + 3 = 3$. Write 3; nothing to carry.

$8 \times 9 = 72$. Write 2; carry 7.

$3 \times 9 + 7 = 34$. Nowhere to carry; write 34.

$$\begin{array}{r} {}^{7}\ {}^{3}\phantom{0} \\ 3804 \\ \times\ 719 \\ \hline 34236 \end{array}$$

**Multiply by 1:**

Either go through and multiply carefully—or realize that $3804 \times 1 = 3804$. Remember to start at the tens' column when recording this row.

$$\begin{array}{r} 3804 \\ \times\ 719 \\ \hline 34236 \\ 3804\phantom{0} \end{array}$$

**Multiply by 7:**

Start at the hundreds' column.

$4 \times 7 = 28$. Write 8; carry 2.

$0 \times 7 + 2 = 2$. Write 2; nothing to carry.

$8 \times 7 = 56$. Write 6; carry 5.

$3 \times 7 + 5 = 26$. Nowhere to carry; write 26.

$$\begin{array}{r} 3804 \\ \times\ 719 \\ \hline 34236 \\ 3804\phantom{0} \\ 26628\phantom{00} \end{array}$$

MULTIPLICATION

**Add everything up**, keeping track of the stacking carefully.

$$
\begin{array}{r}
3804 \\
\times\ 719 \\
\hline
34236 \\
3804\phantom{0} \\
+\ 26628\phantom{00} \\
\hline
2735076
\end{array}
$$

So $3804 \times 719 = 2{,}735{,}076$. Whew! That was a real doozy. You probably won't have to do that very often. But don't you feel strong and powerful now?

# YOUR TURN

Solutions start on page 267.

1. Choose the right answer—**A**, **B**, **C**, or **D**—for each problem.

   (a) $342 \times 100$
   (b) $3420 \times 10$
   (c) $342 \times 1$
   (d) $34{,}200 \times 1$
   (e) $3420 \times 100$
   (f) $34{,}200 \times 10$

      A. 342
      B. 3420
      C. 34,200
      D. 342,000

2. Multiply the following numbers by 5 in your head.

   (a) 14
   (b) 20
   (c) 22
   (d) 17
   (e) 246

3. Find the product of 25 and 7.

4. What is $121 \times 11$?

5. Find $19 \times 19$.

6. Rebecca studies math exactly 43 minutes each Monday, Wednesday, Thursday, and Saturday. How many minutes does Rebecca devote to studying math every week?

7. Multiply:

$$\begin{array}{r} 1234 \\ \times\ 507 \\ \hline \end{array}$$

8. There are 32 students in our after-school remedial-math class. We get homework 3 days a week. Every homework set has exactly 13 problems. If everyone does all the homework, how many math problems does our poor after-school-remedial-math-class teacher have to grade in a week?

# 4

# CHAPTER 4
## DIVISION

### SICK PUPPY

I can't believe I just walked to after-school-remedial-math class with Billy Idol! . . . I mean, Kyle Thomas! He is the coolest creature I have ever met in my short, fast life.

Look at him now, up there at the chalkboard, doing long division, with his cute little butt shaking as he writes, his dirty back pocket peeking out through the hole in his shredded chinos. Oh, wait . . .

Uh-oh. It's getting ugly up there. Kyle doesn't know how to do long division! I can't bear to watch. Those beautiful fingers shouldn't be tentatively holding a broken piece of chalk; they should be holding a guitar pick . . . or me!

"Pssst! How was your walk?" Rebecca is sitting right behind me.

"Great! I helped Kyle with his multiplication tables."

"How . . . um . . . romantic."

"He really is lost up there. He's like a little lost puppy, a little puppy who needs help, who needs love and attention, who needs his tummy rubbed, who needs to be licked clean by his mommy . . ."

"You're a *sick* puppy, Shanon!"

She's right. I have to get a grip. The rest of class was relatively uneventful. Then, as Rebecca and I were leaving class, Kyle and his friend Jake came up to us!

"Hey, Shanon, I wanted to introduce you to my friend Jake. He's my drummer."

"*Your* drummer? I'm the drummer for Johnny London. But I ain't *your* drummer."

"No worries, Jake, didn't mean to get you riled."

"Sorry, I lost my temper," Jake said. He reached in his pants pocket and offered me and Rebecca a half-eaten bar of taffy. "Want some?"

Kyle recoiled. "Jake, mate, you are grossing us out! I wanted to introduce you to Shanon. She's really great at math."

"And so many other things," I said.

"And who might you be?" Kyle asked Rebecca.

"Rebecca," said Rebecca, her eyes narrowing.

"A pleasure to meet you, mantis . . . I mean, *miss*."

"Mantis?" I asked. "Does she look like a bug?"

"No, my fault, I was just thinking of the name for—for our new album," Kyle stuttered. "Anyway, Jake and I wanted to invite you two to our show tonight over at the Alhambra."

"We're there!" I said

"I thought we were studying division tonight!" Rebecca said.

"Maybe we could all study together after the show?" Kyle said.

I almost fainted.

# DIVISION: THE BASICS

Division is closely related to multiplication: you can think of dividing as undoing multiplication.

Most division problems look something like this:

> Shanon has 6 Billy Idol posters. She wants to split them evenly among 3 of the walls of her room (the fourth is taken up by windows). How many posters will she put on each wall?

In math-speak this situation becomes pretty dull:

> How much is 6 posters divided by 3 walls?

Because division is undoing multiplication, this turns out to be the same question as this one:

> How many posters do you have to have on each wall so that when you multiply that number by 3 walls, you get 6 posters?

The answer is 2 posters per wall. You know that because $3 \times 2 = 6$. So

> 6 posters divided by 3 walls = 2 posters per wall.

If you know your times tables, division is going to be easy trudging. I promise.

## TALKING AND WRITING

Division can be expressed in a couple of ways. The most common one is with this funny division sign:

$$30 \div 5 = 6.$$

This is read as "30 divided by 5 equals 6" or "5 **goes into** 30 six times."

In the equation $30 \div 5 = 6$, the number 30 is the **dividend**, 5 is the **divisor**, and 6 is the **quotient**:

$$\overset{\text{dividend}}{30} \div \overset{\text{divisor}}{5} = \overset{\text{quotient}}{6}$$

The term *dividend* is pretty rare, but the other two come up a lot.

Another common way to write down division is used most often when the numbers are large—when you're doing what's called **long division**. More on that later.

$$5\overline{)30}^{\;6}$$

Be careful with this notation: the divisor goes to the left of the funny sign, and the quotient goes on top.

These are the two main ways we'll notate division in this book, but you should be aware of a few others. Sometimes—especially when you move on to algebra—division is written with a slash:

$$30/5 = 6.$$

Or as a fraction, with the bar

either slanted: ${}^{30}/_{5} = 6$ or horizontal: $\dfrac{30}{5} = 6.$

Occasionally, especially on websites, where the ÷ sign is hard to make, division is written with a colon:

$$30 : 5 = 6.$$

### DIVIDING WITH REMAINDER

Let's say Jake and Kyle want to split up 7 pieces of taffy evenly. How many pieces of taffy will each of them get? Well, we want to divide 7 taffy pieces by 2 people—but we can't **divide evenly**. Each person will get 3 pieces of taffy, and 1 piece will be left over. This leftover is called the **remainder**. We can write this down as

$$7 \div 2 = 3, \text{ remainder } 1$$

or as

$$\begin{array}{r} 3 \text{ R1} \\ 2\overline{)7} \end{array}$$

When a division problem leaves no remainder, we say that the divisor **goes into** the number being divided **evenly**. So because $6 \div 3 = 2$ leaves no remainder, 2 goes into 6 evenly. But $7 \div 2$ leaves a remainder, so 2 doesn't go into 7 evenly.

Even though 2 doesn't go into 7 evenly, we can still say that 2 goes into 7 three times, leaving a remainder of 1.

## DIVIDING SMALL NUMBERS

How many times does 7 go into 28? We can rephrase the question like this:

How many times do you have to add 7 to itself to get 28?

or equivalently like this:

By what number should 7 be multiplied to get 28?

Let's think about how to answer both of these questions in turn.



### THE PLODDING-ALONG WAY

How many times do you have to add 7 to itself to get 28?

Well, we could just start adding:

| | | |
|---|---|---|
| One 7 added together: | $7$ | $= 7$ |
| Two 7s added together: | $7 + 7$ | $= 14$ |
| Three 7s added together: | $7 + 7 + 7$ | $= 21$ |
| Four 7s added together: | $7 + 7 + 7 + 7$ | $= 28$ |

Seems like the answer is $4$. So $28 \div 7 = 4$. This way is nice and simple, and it definitely gets the job done—but it sure isn't very efficient.

The numbers $7$, $14$, $21$, and $28$ are all multiples of $7$. That's not a coincidence: multiplication is repeated addition, so adding $7$ to itself will give you multiples of $7$. Fortunately, you've already memorized all the small multiples of $7$ because they're all in the multiplication table.

## USING THE MULTIPLICATION TABLE

By what number should $7$ be multiplied to get $28$?

That's where the multiplication table comes in. (You know it by heart now, right? If not, go back and learn it! It's really crucial.) It tells you that $7 \times 4 = 28$. Well, that's the same thing as $28 \div 7 = 4$. Isn't this great? The multiplication table turns out to be a division table, too.

Let's try another problem for good luck: what's $54$ divided by $9$?

Think back to the multiples of $9$ from the multiplication table. Hopefully you remember that $9 \times 6 = 54$. Which means that $54 \div 9 = 6$.

## REMAINDERS

What's $26$ divided by $4$?

Oh no! The number $26$ isn't one of the multiples of $4$. How to proceed?

**1.** Think of the largest multiple of $4$ less than $26$.

You know that $4 \times 6 = 24$, which is smaller than $26$, but the next multiple is $4 \times 7 = 28$, which is too large. So the largest multiple of $4$ is $24$.

**2.** Divide that multiple by $4$. That's the quotient.

We get $6$.

**3.** To find out the remainder, subtract that largest multiple from $26$.

Since $26 - 24 = 2$, it looks like $2$ is the remainder.

So $26 \div 4 = 6$, remainder $2$.

Let's try just one more example. What's $52$ divided by $7$?

The number $7$ doesn't go into $52$ evenly, so we think of the largest multiple of $7$ less than $52$. That seems to be $49$ ($7 \times 7 = 49$, but $7 \times 8 = 56$). So $7$ goes into $52$ seven times, leaving a remainder of $52 - 49 = 3$:

$$52 \div 7 = 7, \text{ remainder } 3.$$

## KEEP THE REMAINDER SMALL!

This is important—so important that I'm going to put it in a box:

| **The remainder must be smaller than the number you're dividing by.** |
|---|

If it's not, you made a mistake. Go back and fix it.

Let's see how that works. Suppose we messed up and thought that the largest multiple of $7$ less than $52$ is $42$. After all, $7 \times 6 = 42$, so it's not such a crazy thought. So we conclude that the quotient is $6$ and the remainder is $52 - 42 = 10$.

Oops. The remainder $10$ is bigger than the divisor $7$. That's no good. It means that we didn't find the largest multiple of $7$: there's another, larger multiple that works better—which we already knew, of course.

## CHECKING YOUR WORK

As I've been saying, division undoes multiplication. Conveniently, multiplication also undoes division. So you can check division problems with multiplication.

Say you determine that $63 \div 7 = 9$. You can check your work by multiplying the quotient by the divisor to see if you get the original number you divided back. In this case, $9 \times 7 = 63$, so we're good.

You can even check your work if you got a remainder. Suppose that you divided and got $34 \div 4 = 8$, remainder $2$. Check your work by multiplying the quotient by the divisor and adding the remainder. You should get the number you started with:

$$\text{original number} = \text{divisor} \times \text{quotient} + \text{remainder}.$$

In this case, $8 \times 4 + 2 = 34$, which is good news.

## (NEVER EVER) DIVIDING BY ZERO

Another extremely important fact:

| **You cannot divide by zero. Ever.** |
|---|

Why not?

Essentially, to solve something like $4 \div 0$, you have to find the number that, when multiplied by $0$, gives $4$. That's never going to happen: $0$ times any other number is still $0$.

But wait, you might say. Maybe I can't divide $4$ by $0$ because it's not a multiple of $0$. But $0$ is a multiple of $0$. So I can do $0 \div 0$, right?

Wrong. To divide $0$ by $0$, you'd have to find the number that, when multiplied by $0$, gives $0$. But *any* number does that: $1 \times 0 = 0$, and $13 \times 0 = 0$, and even $0 \times 0 = 0$. There's no unique number that works. So there's no answer.

I'm just going to say it again so you'll get annoyed and remember it. You can never, *ever*, *ever* divide by zero.

So what should you write down if someone (me, for example) tries to be sneaky and get you to answer something like this:

$$12 \div 0 = ?$$

Write down "can't do" or "no answer" or "impossible" or, if you want to be fancy, "DNE" (for "does not exist"). 'Cause you can't do it; there's no answer; it's impossible.

# LONG DIVISION

When the number being divided is too large—larger than the numbers in from the multiplication table—we resort to a procedure called **long division**, in which we find the quotient digit by digit.

## FIRST EXAMPLE

$87 \div 3 = ?$

First, **set this up** with the funny long-division sign. Make sure to leave some room under and to the right of the $87$.

$$3\overline{)87}$$

**First digit:** Next, take the first digit of the number being divided—in this case, that's $8$—and divide it by $3$. You know that $3$ goes into $8$ two times. Great: $2$ is the first digit of the quotient. Write down $2$ on top, where the quotient will go. It should go directly over the $8$. Don't worry about the remainder for the moment.

$$\begin{array}{r} 2\phantom{7} \\ 3\overline{)87} \end{array}$$

Now, multiply $3$ by $2$ and write the result just under the $8$:

$$
\begin{array}{r}
2\phantom{7} \\
3\overline{)87} \\
6\phantom{7}
\end{array}
$$

Draw a line under the $6$, subtract $6$ from $8$, and write the difference under the line. This is the remainder when $8$ is divided by $3$.

$$
\begin{array}{r}
2\phantom{7} \\
3\overline{)87} \\
\underline{-6\phantom{7}} \\
2\phantom{7}
\end{array}
$$

**Bring down** the $7$—the next digit from the number being divided.

$$
\begin{array}{r}
2\phantom{7} \\
3\overline{)87} \\
\underline{-6\phantom{7}} \\
27
\end{array}
$$

**Second digit:** You've created a $27$. Divide it by $3$ to get $9$: this is the next digit in the quotient. Write it in, above the $7$ in $87$.

$$
\begin{array}{r}
29 \\
3\overline{)87} \\
\underline{-6\phantom{7}} \\
27
\end{array}
$$

Now, repeat what you did with the previous digit: multiply $3$ by $9$, write the result under the $27$, and subtract.

$$
\begin{array}{r}
29 \\
3\overline{)87} \\
\underline{-6\phantom{7}} \\
27 \\
\underline{-27} \\
0
\end{array}
$$

There are no more digits to bring down, so we're done: $87 \div 3 = 29$. Because the last difference is $0$, there's no remainder.

## REMAINDERS AND ALIGNMENT

That wasn't so bad, was it? Let's do another example.

Divide $2597$ by $4$.

**Set up:**

$$4\overline{)2597}$$

**Find the first digit of the quotient:** The first digit of $2597$ (which is $2$) is too small to be divided by $4$, so instead take the first *two* digits as a two-digit number. The divisor $4$ goes into $25$ six times. Write down $6$ just over the last digit of $25$. There's space for two more digits in the quotient, so let's just mark them with dots. This way we won't forget that the quotient will be a three-digit number.

$$\begin{array}{r} 6\cdot\cdot \\ 4\overline{)2597} \end{array}$$

Multiply the divisor by the first digit in the quotient ($4 \times 6 = 24$), subtract ($25 - 24 = 1$), and bring down the next digit (place the $9$ next to the $1$).

$$\begin{array}{r} 6\cdot\cdot\; \\ 4\overline{)2597} \\ \underline{-24\phantom{97}} \\ 19\phantom{7} \end{array}$$

**Do it all all over again:** $19$ divided by $4$ gives $4$, which is the next digit of the quotient. Multiply $4$ by $4$, write under the $19$, subtract. Bring down the $7$.

$$\begin{array}{r} 64\cdot \\ 4\overline{)2597} \\ \underline{-24\phantom{97}} \\ 19\phantom{7} \\ \underline{-16\phantom{7}} \\ 37 \end{array}$$

**And one more time:** 37 divided by 4 is 9. Write it down as the last digit of the quotient. Compute $4 \times 9 = 36$; write it down under 37. Subtract.

$$
\begin{array}{r}
649 \\
4\overline{)2597} \\
-24\phantom{97} \\
\hline
19\phantom{7} \\
-16\phantom{7} \\
\hline
37 \\
-36 \\
\hline
1
\end{array}
$$

There are no more digits to bring down, so we're done. The leftover 1 is the remainder in the division: $2597 \div 4 = 649$, remainder 1. To make sure the remainder doesn't get forgotten, it's sometimes rewritten next to the quotient:

$$
\begin{array}{r}
649 \text{ R1} \\
4\overline{)2597}\phantom{\text{ R1}} \\
-24\phantom{97}\phantom{\text{ R1}} \\
\hline
19\phantom{7}\phantom{\text{ R1}} \\
-16\phantom{7}\phantom{\text{ R1}} \\
\hline
37\phantom{\text{ R1}} \\
-36\phantom{\text{ R1}} \\
\hline
1\phantom{\text{ R1}}
\end{array}
$$

You can check your work by multiplying 649 by 4 and adding 1. You should get 2597.

## PESKY ZEROS

You must be sick of this by now, but let's just do one more quick example:

$$8\overline{)4872}$$

**First digit of the quotient:** $4$ is too small, so start with $48$. The divisor $8$ goes into $48$ six times. Write down $6$ over the $8$. Two more digit-spaces are left, so the quotient will be a three-digit number. Multiply $6 \times 8$ and subtract.

$$\begin{array}{r} 6\cdot\cdot \\ 8\overline{)4872} \\ -48\phantom{72} \\ \hline 0\phantom{72} \end{array}$$

The difference is a zero, but don't let that faze you. Bring down the $7$ as usual.

$$\begin{array}{r} 6\cdot\cdot \\ 8\overline{)4872} \\ -48\phantom{72} \\ \hline 07\phantom{2} \end{array}$$

**Second digit of the quotient:** $7$ is too small to be divided by $8$. So we bring down the $2$ as well. But to mark the fact that $7$ is too small, and to make sure that all of our digits line up properly, we write down $0$ as the next digit of the quotient. Like this:

$$\begin{array}{r} 60\cdot \\ 8\overline{)4872} \\ -48\phantom{72} \\ \hline 72 \end{array}$$

**Third digit of the quotient:** $72$ is not too small, so we divide normally: $8$ goes into $72$ nine times. Write down $9$; multiply $8 \times 9 = 72$ and subtract.

$$\begin{array}{r} 609 \\ 8\overline{)4872} \\ -48\phantom{72} \\ \hline 72 \\ -72 \\ \hline 0 \end{array}$$

Done: $4872 \div 8 = 609$. Beautiful.

# TWO-DIGIT DIVISORS

That's it! No more new stuff to learn.

Well, almost. I just want to go through a couple of examples of dividing by a two-digit number. The good part is that it works exactly the same way. The bad part is that since we don't memorize multiples of two-digit numbers, it's sort of hard to guess the digits of the quotient. We have to do some yucky trial and error.

Enough talk; let's just dig right in.

$$13\overline{)1066}$$

**First digit of the quotient:** Clearly, $1$ is too small. But it looks like $10$ is *also* too small. We're going to have to get the first digit from dividing $106$ by $13$. And the bitter truth is that there's no good way to know how many times $13$ goes into $106$. You just have to guess. See what I mean about yucky?

How do you make the guess? By hook or by crook, as they say. For example, I know that $10$ goes into $106$ ten times, because $10 \times 10 = 100$. And $13$ is a little bigger, so let's guess that $13$ goes into $106$ nine times.

Before you write down $9$ in the quotient, check it by multiplying. (You'd have to multiply eventually, so it's not actually extra work.)

$$\begin{array}{r} {}^{2}\phantom{3} \\ 13 \\ \times\, 9 \\ \hline 117 \end{array}$$

Hmm. The product that we got—$117$—is bigger than $106$. So it looks like $9$ isn't going to work. It's too big. But $117$ is pretty close to $106$, so maybe $8$ will work. Let's check it:

$$\begin{array}{r} {}^{2}\phantom{3} \\ 13 \\ \times\, 8 \\ \hline 104 \end{array}$$

Excellent: $104$ is just a little smaller than $106$. So $8$ is the right digit. Let's write it down; it goes just above the last digit of $106$. One more digit left, so the quotient will have two digits.

$$\begin{array}{r} 8\,\cdot \\ 13\overline{)1066} \end{array}$$

You already know that $13 \times 8 = 104$, so write that down, subtract, and bring down the $6$.

$$\begin{array}{r} 8\,\cdot \\ 13\overline{)1066} \\ -104 \\ \hline 26 \end{array}$$

**Second digit of the quotient:** How many times does $13$ go into $26$? Two times is a pretty good guess. Let's check it:

$$\begin{array}{r} 13 \\ \times 2 \\ \hline 26 \end{array}$$

Looks good. Write it down and complete the problem.

$$\begin{array}{r} 82 \\ 13\overline{)1066} \\ -104 \\ \hline 26 \\ -26 \\ \hline 0 \end{array}$$

## DIVIDING BY 10

One super-last, super-easy thing. Remember how multiplying by $10$ is the same thing as adding on a zero at the end? (Take another look at Chapter 3 if that doesn't ring a bell.) So $23 \times 10 = 230$, and so on.

Since division and multiplication are so closely related, there's an analogous trick for division. If a number ends in a $0$ digit, then dividing by $10$ is the same thing as lopping off that last $0$.

Take a look:

$$\begin{aligned} 50 &\div 10 = 5 \\ 230 &\div 10 = 23 \\ 14{,}670 &\div 10 = 1467 \\ 9200 &\div 10 = 920 \\ 1{,}000{,}000 &\div 10 = 100{,}000 \end{aligned}$$

If a number doesn't end with $0$, then it isn't evenly divisible by $10$. But you can still divide: chop off that last digit and call it the remainder.

So

$$235 \div 10 = 23, \text{ remainder } 5,$$

and

$$3001 \div 10 = 300, \text{ remainder } 1.$$

You're done! Now go do the exercises to make sure that you've got it.

# YOUR TURN

Solutions start on page 270.

1. Divide with remainder, in your head.

   (a) $22 \div 3$

   (b) $35 \div 4$

   (c) $34 \div 5$

   (d) $51 \div 6$

   (e) $30 \div 7$

   (f) $60 \div 8$

   (g) $80 \div 9$

2. I asked Rebecca to divide $57$ by $11$. Here's what she got:

   $$57 \div 11 = 4, \text{ remainder } 13.$$

   "That looks really fishy," I said. So she checked her division:

   $$\begin{array}{lllll} \text{divisor} & \times \text{ quotient} & + \text{ remainder} & = \text{original number} \\ 11 & \times \quad 4 & + 13 & = 55. \end{array}$$

   So who's right, me or Rebecca? And if I'm right, why does the check work out?

3. Find the quotient and the remainder when $467$ is divided by $3$.

4. Divide $741$ by $7$.

5. Estimate in your head. How many times . . .

   (a) does $10$ go into $567$?

   (b) does $20$ go into $160$?

   (c) does $30$ go into $230$?

   (d) does $41$ go into $355$?

   (e) does $89$ go into $640$?

(f) does $24$ go into $204$?

(g) does $57$ go into $277$?

6. Last Thursday I arrived at the Finer Diner with a serious hankering for sweet-potato fries. Each order of sweet-potato fries costs $\$4$, and I had $\$50$ on me. How many orders of sweet-potato fries can I afford to chomp down? (For the record, $\$50$ is *not* my usual pocket change, I assure you. My allowance is pathetic. But I'd just seen my grandmother the night before.)

7. Find the quotient and the remainder: $11\overline{)787}$.

8. Perform the division: $10{,}353 \div 17$.

9. Jake divided a number by $7$ and got $12$ with a remainder of $3$. What was the number Jake was molesting so?

10. Becca divided $168$ by some number and got $6$ evenly. What was her divisor?

11. What's wrong with saying $4 \div 0 = 0$, remainder $4$?

# 5

# CHAPTER 5
# OPERATIONS AND PROPERTIES

## MATH MATE

The Johnny London show at the Alhambra totally rocked! Kyle is definitely the best English singer-slash-songwriter-slash-guitar-player in Maplewood. Those big lips, that fine fanny, that I'm-bored-with-life way he leans against the amps when he's not singing . . . That glorious ennui!

And now we're back at the Finer Diner, studying math. Life really doesn't get any better than this.

"What can I get you kids?" the waitress asked.

"Pepsi," said Rebecca.

"Small salad with a side order of taffy," said Jake.

"Peanuts and cheese," I said.

"Pint of beer," said Kyle.

"Can I see some ID?" said the waitress. He showed her his passport.

"Be right back," said the waitress.

I just stared at Kyle.

"He's twenty-one," said Jake.

I stared some more.

"Got a late start in life, mate." Kyle smiled.

Mate, yes. I *really* like it when he calls me that. So let me get this straight: I'm closing in on a boyfriend who's twenty-one and British and crappy at math and plays for the best band in Maplewood? It really, really, really doesn't get any better than this.

"So are we going to study math or what?" Rebecca asked. "I have places to be."

"Like where?" I asked.

"Like anywhere but here," Rebecca said.

"Maybe we do a little studying, and then maybe we do something else?" Kyle asked.

"Fine," said Rebecca. "First let's review long division again, then I need help with some of this order-of-operations nonsense. And then I have to . . . go."

"Where are you going?" I protested.

"I have to attend to some very important business."

See what I mean about Rebecca? She's very mysterious.

# ORDER OF OPERATIONS

Now that you're a master adder, subtracter, multiplier, and divider, we can start putting some of this stuff together.

By the way, addition, subtraction, multiplication, and division are collectively known as the four basic **operations**—they're all procedures we do to numbers.

## DOES ORDER MATTER?

Let's ask Jake and Kyle to solve the same problem: $7 + 8 - 3$.

There are two operations here: addition and subtraction. They're going to have to pick which one to do first.

| Kyle adds, then subtracts | Jake subtracts, then adds |
|---|---|
| $7 + 8 - 3 =$ | $7 + 8 - 3 =$ |
| $15 \quad - 3 = 12$ | $7 + \quad 5 \quad = 12$ |

So far, so good. They got the same answer. It looks like it doesn't matter which operation you do first, at least if you're choosing between addition and subtraction.

But let's ask Jake and Kyle to do another problem: $4 + 6 \times 2$.

This one also has two operations: addition and multiplication. Easy enough. Jake and Kyle don't want to complicate anything, so they do the same thing.

| Kyle adds, then multiplies | Jake multiplies, then adds |
|---|---|
| $4 + 6 \times 2 =$ | $4 + 6 \times 2 =$ |
| $10 \quad \times 2 = 20$ | $4 + \quad 12 \quad = 16$ |

Uh-oh. They got different answers. What does this mean? Who's right?

## MULTIPLY AND DIVIDE FIRST

What this means is that the order in which you do operations matters. It won't matter *every single* time—you saw that it made no difference in the first example—but it will matter very often.

To eliminate confusion, people have adopted some conventions. If there's nothing special in the equation—no parentheses or other order-defying things—multiplication and division are done before addition and subtraction. So your plan of attack is as follows:

1. Working from left to right, do any **multiplication** or **division**.
2. Working from left to right, do any **addition** or **subtraction**.

So who was right about $4 + 6 \times 2$? Jake: he multiplied first. (Kyle has a lot of trouble with this stuff. Helpless . . . but *so* adorable.)

## BASIC EXAMPLES

Let's run through a couple of examples.

First one: $20 - 8 \div 4 + 7$

First multiply and divide, left to right; then add and subtract, left to right. So we'll end up doing the operations in this order:

$$20 \overset{②}{-} 8 \overset{①}{\div} 4 \overset{③}{+} 7.$$

**Multiply & divide:**

$$(1) \quad 20 - 8 \div 4 + 7 =$$
$$20 - 2 + 7 =$$

**Add & subtract:**

$$(2) \quad 20 - 2 + 7 =$$
$$(3) \quad 18 + 7 = 25$$

And another nastier one: $5 + 32 \div 4 \times 3 - 2 \times 7$

We do the operations in the following order:

$$5 \overset{④}{+} 32 \overset{①}{\div} 4 \overset{②}{\times} 3 \overset{⑤}{-} 2 \overset{③}{\times} 7.$$

**Multiply & divide:**

$$(1)\ 5 + 32 \div 4 \times 3 - 2 \times 7 =$$
$$(2)\ 5 + 8 \times 3 - 2 \times 7 =$$
$$(3)\ 5 + 24 - 2 \times 7 =$$
$$5 + 24 - 14$$

**Add & subtract**:

$$(4)\ 5 + 24 - 14 =$$
$$(5)\ 29 - 14 = 15$$

# PARENTHESES

So multiplication and division normally come before addition and subtraction. But what if we *need* to add first?

For example, say Jake and Kyle both need new guitar picks. Jake needs $4$ picks, and Kyle needs $6$ picks. Each pick costs $\$2$. How much does Johnny London need to spend on picks?

In order to get the total pick cost, Jake and Kyle have to first add Jake's picks and Kyle's picks together, and then multiply by $\$2$ per pick to get $\$20$. Something like $4 + 6 \times 2$ doesn't work: the conventional order of operations insists that multiplication be done before addition. So people use **parentheses** to group operations that have to be done first. Like so:

$$(4 + 6) \times 2 = 20.$$

Nice, huh? Parentheses trump over the normal order of operations.

By the way, the word *parentheses*—pronounced *pah-REN-thuh-SEEZ*—is plural. It refers to a set of two: ( and ). Either one symbol alone is called a **parenthesis**, pronounced *pah-REN-thuh-sis*.

Parentheses are terrifically useful when you want to perform an addition or subtraction before a multiplication or division, reversing the regular order of operations. But they can also come in handy in other situations, say, with two divisions or two subtractions.

For example, compare $24 \div 4 \div 2$ and $24 \div (4 \div 2)$.

$$24 \div 4 \div 2 = 6 \div 2 = 20 \qquad 24 \div (4 \div 2) = 24 \div 2 = 12$$

Parentheses in a mathematical expression are always saying, *Do this stuff first.*

OPERATIONS & PROPERTIES

## OPERATIONS WITHIN PARENTHESES

Parentheses can enclose any number of operations. Within a parenthesis pair, do the operations in the regular order: multiplication and division, then addition and subtraction.

Let's try an example: $4 - 6 \div (23 - 4 \times 5)$

First focus on the operations within the parentheses. There's a multiplication and a subtraction; the multiplication comes first. Outside the parentheses, division comes before subtraction, so the overall order is as follows:

$$\overset{④}{4 -} \overset{③}{6 \div} (\overset{②}{23 -} \overset{①}{4 \times} 5).$$

**In parentheses, multiplication and division:**

(1) $4 - 6 \div (23 - 4 \times 5) =$

**In parentheses, addition and subtraction:**

(2) $4 - 6 \div (23 - 20) =$

**Outside, multiplication and division:**

(3) $4 - 6 \div (3) =$

**Outside, addition and subtraction:**

(4) $4 - 2 = 2$

## NESTED PARENTHESES

Parentheses that appear inside other parentheses are called **nested**. (They sit inside the "nest" of another set of parentheses.) As I've mentioned, operations within parentheses also follow order-of-operation rules—parentheses, then multiplication and division, then addition and multiplication. So nested parentheses are evaluated first.

This means that in $2 \times (15 - (4 + 5) \div 3)$, the sum $4 + 5$ is evaluated first. The operations should be done in this order:

$$\overset{④}{2 \times} \big(\overset{③}{15 -} (\overset{①}{4 +} 5) \overset{②}{\div} 3\big).$$

You should get that $2 \times (15 - (4 + 5) \div 3) = 24$.

### PLEASE EXCUSE MY DEAR AUNT SALLY

You may have already heard this phrase. It's a "mnemonic"—a memory trick to help you remember the correct order of operations. The first letter of each word stands for an element in the order rules.

| | |
|---|---|
| **P** | **P**arentheses |
| **E** | **E**xponents (Chapter 12) |
| **M, D** | **M**ultiplication and **D**ivision |
| **A, S** | **A**ddition and **S**ubtraction |

You can remember the acronym **PEMDAS** with the phrase "Please excuse my dear Aunt Sally."

## PROPERTIES OF OPERATIONS

The four operations have various nice properties, which can help make computation easier. Some of these we've already talked about; others might be new. Even if new, all of these will likely seem obvious—except perhaps the distributive property at the very end. But here they all are, together.

One last thing: unless your teacher is a stickler for such things, don't worry about memorizing the names of these properties. It's much more important to be able to use them than to name them.

### ADDITION: ORDER DOESN'T MATTER

We talked about this one in Chapter 1. It says that you can flip the order of the numbers that you're adding without changing the sum. So

$$5 + 8 = 8 + 5 \quad \text{and} \quad 109 + 13 = 13 + 109.$$

We say that *addition is commutative*. This property is called the **commutative property of addition**.

### ADDITION: REPARENTHESIZING

You've probably already noticed that this is true: if you have two plus signs in a row, you can do either one first. Essentially, you can reparenthesize, like this:

$$(5 + 2) + 3 = 5 + (2 + 3).$$

Check that this is true: $(5 + 2) + 3 = 7 + 3 = 10$.

And $5 + (2 + 3) = 5 + 5 = 10$.

Beautiful. This rule is helpful if you're calculating something like $(74 + 51) + 9$. Normally, you'd have to add $74 + 51$ first–which is fine, but a little annoying. Instead, you can rearrange the parentheses to compute $74 + (51 + 9)$. Now, the first calculation is $51 + 9$, which is much easier: you get $60$, which is a round number.

We say that *addition is associative*–we can change the groupings (or "associations") of the numbers that we add. This property is called the **associative property of addition**.

Together with the previous property, this means that addition of several numbers together can be done in any order whatsoever–which is convenient.

## ADDING OR SUBTRACTING ZERO

Adding or subtracting zero doesn't do a thing.

So

$$4 + 0 = 4$$

and

$$0 + 59 = 59$$

and

$$10{,}933 - 0 = 10{,}933.$$

This is true even when you haven't computed the number that you're adding yet. For example,

$$(78 + 89) + 0 = 78 + 89.$$

## MULTIPLICATION: ORDER DOESN'T MATTER

We talked about this one as well, in Chapter 3. It says that you can switch the order of the numbers that you're multiplying without affecting the product. So

$$5 \times 8 = 8 \times 5 \quad \text{and} \quad 19 \times 81 = 81 \times 19.$$

We say that *multiplication is commutative*, just like addition. This is the **commutative property of multiplication**.

## MULTIPLICATION: REPARENTHESIZING

This is another multiplicative property that parallels an additive property: if you have two times signs in a row, you can do either one first. Essentially, you can reparenthesize, like this:

$$(7 \times 2) \times 8 = 7 \times (2 \times 8).$$

Check it out: $(7 \times 2) \times 8 = 14 \times 8 = 112$. And $7 \times (2 \times 8) = 7 \times 16 = 112$. Looks like it works.

Again, this rule is helpful because some numbers are easier to multiply than others. For example, take $(71 \times 5) \times 2$. Normally, you'd compute $71 \times 5$ first, then multiply the nasty product by $2$. In short, a bother. Instead, rearrange the parentheses to make $71 \times (5 \times 2)$. Now the first computation is much easier: $5 \times 2 = 10$. And because multiplying by $10$ is super-easy—just tack on a zero at the end—the whole problem becomes very simple: $71 \times 10 = 710$.

The name for this property is the same as for the corresponding additive property. We say that *multiplication is associative*, and the property is called the **associative property of multiplication**.

And again, the two multiplicative-order properties together mean that the product of several numbers can be computed in any order.

## MULTIPLYING AND DIVIDING BY ONE

Multiplying or dividing by one doesn't do much.

So

$$6 \times 1 = 6$$

and

$$1 \times 57 = 57$$

and

$$89 \div 1 = 89.$$

This is true even when you're don't know yet exactly what quantity you're multiplying. For example,

$$1 \times (56 - 63 \div 9) = 56 - 63 \div 9.$$

## MULTIPLYING BY ZERO

On the other hand, multiplying by zero is an apocalyptic event. Everything gets annihilated and becomes zero:

$$5 \times 0 = 0$$

and

$$0 \times 189 = 189.$$

This fact can save you a lot of needless work. For example,

$$(256 - 78 \div 3 + 67{,}888) \times 0 = 0.$$

Don't even bother working out that dreadful mess in parentheses!

## THE DISTRIBUTIVE PROPERTY

I've saved the best for last. This is the only property that you may not have a feel for. It's best illustrated with an example:

$$3 \times (5 + 4) = 3 \times 5 + 3 \times 4.$$

Did you catch that? Instead of multiplying the $3$ by the sum of $5$ and $4$, we can multiply $3$ by $5$ and by $4$ separately and then add the products together. We open up the parentheses and **distribute** the $3$ over the sum:

$$\textcircled{3} \times (5 + 4) = \textcircled{3} \times 5 + \textcircled{3} \times 4.$$

Rather than multiplying by a sum, you add two products. Let's check: $3 \times (5 + 4) = 3 \times 9 = 27$. On the other hand, $3 \times 5 + 3 \times 4 = 15 + 12$, which is equal to $27$.

Formally, we can say that "multiplication distributes over addition." But most people just call this the **distributive property**: the number multiplied by a sum gets *distributed* to each part that makes up the sum.

You can also distribute if the sum comes first:

$$(7 + 4) \times 6 = 7 \times 6 + 4 \times 6.$$

And you can distribute multiplication over subtraction in the same way. Just don't lose the minus sign:

$$2 \times (10 - 5) = 2 \times 10 - 2 \times 5$$

and

$$(7 - 3) \times 4 = 7 \times 4 - 3 \times 4.$$

The distributive property is helpful in two ways.

**Way one:** Sometimes you have to multiply by a difficult number. For example, $(100 + 3) \times 5$ is much easier to compute by distributing first:

$$100 \times 5 + 3 \times 5 = 500 + 15 = 515.$$

**Way two:** On the other hand, sometimes you can apply the distributive property in reverse to cut down on the number of computations that you have to do. For example, $6 \times 7 + 3 \times 7$ involves three separate operations. But if you notice that both numbers are being multiplied by $7$, you can collapse the problem into $(6 + 3) \times 7$, which simplifies into $9 \times 7$, or $63$. You've saved yourself one whole multiplicative step.

## DISTRIBUTING THE MINUS SIGN

There is a little special case of the distributive property that we should review, very quickly. It doesn't appear to involve multiplication, but you should still think of this as the distributive property.

If you're subtracting a sum, like so,

$$40 - (3 + 4)$$

you can open up the parentheses—but the minus sign gets repeated where the plus sign used to be:

$$40 - (3 + 4) = 40 - 3 - 4.$$

**Check:** $40 - (3 + 4) = 40 - 7 = 33$. On the other hand, $40 - 3 - 4 = 37 - 4 = 33$. This works because you want to subtract the sum of $3$ and $4$ from $40$, which is the same thing as subtracting both $3$ and $4$, or first $3$ and then $4$.

If you're subtracting a difference, the minus sign in the parentheses gets changed to a plus when the parentheses open up:

$$20 - (5 - 3) = 20 - 5 + 3.$$

**Check:** $20 - (5 - 3) = 20 - 2 = 18$. On the other hand, $20 - 5 + 3 = 15 + 3 = 18$. This works because when you subtract something *less* than $5$ from $20$, you're going to get something *more* than $20 - 5$. You have to add to make sure that you don't subtract too much.

In general, when subtracting a sum or difference, you can open up the parentheses if you flip all the signs inside them: plus to minus and minus to plus.

$$10 - (5 - 6 + 7 - 4) = 10 - 5 + 6 - 7 + 4.$$

This flipping is called **distributing the minus sign**.

# YOUR TURN

Solutions start on page 273.

1. Simplify each expression.

   (a) $28 \div 4 - 2 \times 3$

   (b) $45 + 2 \times 7 - 12$

   (c) $34 \div (1 + 2 \times 2 \times 4) \times (56 \div 7)$

   (d) $13 + (15 - 14) \times 13 - 25$

   (e) $(18 \times (12 \div 3) - (7 - 1)) \div (12 - (6 + 3))$

2. Do each problem in your head.

   (a) $34 \times 1$

   (b) $46 \times 0$

   (c) $12 \div 0$

   (d) $17 \div 1$

   (e) $0 \times (12 \div 3 \times 8 - 56)$

   (f) $9 \times 8 \times 7 \times 6 \div 4 \div 3 \div 2 \div 1 \times 0$

3. Do each problem in your head. If you used a property of operations, name it.

   (a) $234 \times 7 + 234 \times 3$

   (b) $122 + 37 - 122$

   (c) $145 \times 2 \div 145$

   (d) $12 \times 789 - 2 \times 789$

   (e) $(100 + 1) \times 73$

   (f) $(34 \times 5) \times 2$

   (g) $1 + (999 + 678)$

4. Rewrite each expression to undo the parentheses. Do not simplify further.

(a) $(9 + 7) \times 2$

(b) $(9 - 7) \times 2$

(c) $2 \times (9 + 7)$

(d) $2 \times (9 - 7)$

5. Rewrite each expression to undo the parentheses. Do not simplify further.

(a) $11 + (9 + 2)$

(b) $22 + (8 - 3)$

(c) $33 - (7 + 4)$

(d) $44 - (6 - 5)$

6. Without doing any calculations, determine which of the following expressions name the same value as

$$1234 + 5678 \times 4567.$$

(a) $5678 + 1234 \times 4567$

(b) $4567 \times 5678 + 1234$

(c) $4567 \times 1234 + 5678$

(d) $(1234 + 5678) \times 4567$

(e) $1234 + (5678 \times 4567)$

(f) $1234 + 0 + 4567 \times 5678$

(g) $1234 \times 0 + 1 \times 4567 \times 5678$

# 6

# CHAPTER 6
## FACTORING



### THE MAPLES HAVE EYES

We finished going over the distributive property at the Finer Diner, and Rebecca went home. It was just me and Kyle! Oh, and Jake.

"Could I get this taffy to go?" Jake asked. The waitress raised her eyebrows, but then left to pack a doggie bag.

"I have a recording of our last show at the Aragorn, mate," Kyle said to me. "It's back at home. You want to hear it?"

"You played at the Aragorn?!" I exclaimed.

"We opened for Black Star," Jake said.

"Sounds like a plan!" I said. Kyle paid for everything—left a nice tip, too—and we stepped out into the cool evening air. Jake stopped to tie his shoe. Kyle and I walked ahead a few feet. That's when he dropped the bomb on me.

"I really like you, Shanon," he said. For that second I was the happiest girl in the world. Then my world got *rocked.*

Out of nowhere, a black-suited ninja assassin flew on top of Kyle and knocked him to the ground, applying an illegal sleeper hold. Jake dropped his taffy bag and rushed to defend his lead singer but was stopped dead in his tracks by a poison-tipped dart shot by the ninja assassin's partner, who dropped like a spider from one of Maplewood's many maple trees.

I backed into a big flowering pricker bush, screaming my head off. That's when the ninja who had incapacitated Kyle ripped his mask off.

"Dad!" I exclaimed. Then the other ninja did the same.

"Dad's beautiful blond secretary!" I screamed. Then I fainted.

* * * * *

When I woke up, I was in a cave equipped with computers and the latest in global-positioning satellite technology. My dad and his beautiful blond secretary were tracking a red blip as it blipped across one of the ten flat-screen TV panels suspended from the drippy rock ceiling.

Someone put a cold cloth on my head. I looked up.

"Mom!" I yelled. Seeing that I was awake, Dad and his beautiful blond secretary interrupted their blip-watching and came over.

"Are you okay, honey?" Mom asked.

"What is going on here, Mother?" I asked. "And what have Dad and his beautiful blond secretary done to my new boyfriend, Kyle?"

"He's over in that cramped jail cell, sweetie pie," Mom said. And so he was, shackled to the bars along with Jake. Neither of them was moving. I wanted to cry.

"Is this because he's older than me?!" I asked. "I know parents are supposed to be overprotective, but this is ridiculous."

"No, honey," Dad said. "It's because he's a rogue spy bent on world domination. Your mother and I and my beautiful blond secretary are all undercover operatives working to foil his fiendish plot."

"You'll have to stay here tonight while we finalize our plan," added Mom.

"But I promised Rebecca I'd help her with factoring tomorrow morning!" I protested.

"Write her an email," Mom said, handing me a laptop. "But mention none of this."

* * * * *

FROM: Shanon Fletcher [shanon4@britrockmail.com]
TO: Rebecca Romaine [rr009@blackmantis.net]
SUBJECT: Factoring (won't be around tomorrow morning; see you in school)

Sometimes we care more about whether it's possible to divide evenly, leaving no remainder, than about the result of division. This chapter will set us up for working with fractions in Chapters 7, 8, and 9. Also, it's fun—relatively speaking, of course.

# FACTORS

Let's look at a multiplication equation again:

$$4 \times 9 = 36.$$

You already know that 36 is a **multiple** of both 4 and 9. Also, 36 is **divisible** by both 4 and 9. That's because both 4 and 9 **go into** 36 **evenly**. Both 4 and 9 are **factors** of 36. Less often, they are called **divisors** of 36. So any number is divisible by all of its divisors. Get it?

We can think about **factoring** $36$. That just means that we think of a way of writing $36$ as a product of two numbers. So $36$ **factors** as $4 \times 9$. It also factors as $12 \times 3$. A number can factor in more than one way.

To recap, all of the following things are true:

- 4 is a factor of 36
- 9 is a divisor of 36
- 4 goes into 36 evenly
- 36 is divisible by 9
- 36 is a multiple of 4
- 36 factors as the product of 4 and 9

## FINDING ALL THE FACTORS

Occasionally we need to find all the factors of a particular number, as a list. For example, what are all the factors of $10$?

To find factors, try to factor $10$ in all possible ways. For example, $10 = 2 \times 5$ and $10 = 1 \times 10$. To make sure that you find all such **factorizations**, keep incrementing the first factor, one by one:

| Possible factor | Factorization |
|---|---|
| $1$ | $10 = 1 \times 10$ |
| $2$ | $10 = 2 \times 5$ |
| $3$ | Not divisible by $3$ |
| $4$ | Not divisible by $4$ |
| $5$ | $10 = 5 \times 2$ |
| $6$ | Not divisible by $6$ |
| $7$ | Not divisible by $7$ |
| $8$ | Not divisible by $8$ |
| $9$ | Not divisible by $9$ |
| $10$ | $10 = 10 \times 1$ |

So $10$ has four factors: $1$, $2$, $5$, and $10$.

The list above includes both $10 = 2 \times 5$ and $10 = 5 \times 2$. But we already know that order doesn't matter in multiplication, so listing both factorizations is a waste of time. To save time, stop listing factorizations when the second factor starts being bigger than the first.

Take a look at how this works with the factors of $18$:

$$18 = 1 \times 18$$
$$18 = 2 \times 9$$
$$18 = 3 \times 6$$

We can stop here. The next factorization is $18 = 6 \times 3$, but we already know both of those factors from $18 = 3 \times 6$. So we're done. The factors of $18$ are $1$, $2$, $3$, $6$, $9$, and $18$.

I want to do just one more: $60$. I like it because it has so many factors!

$$60 = 1 \times 60$$
$$60 = 2 \times 30$$
$$60 = 3 \times 20$$
$$60 = 4 \times 15$$
$$60 = 5 \times 12$$
$$60 = 6 \times 10$$

Since $60$ is not divisible by $7$, $8$, or $9$, we can stop. The next factor is $10$, but it's already listed. Isn't this neat? The number $60$ has twelve factors and is divisible by *every number* between $1$ and $6$!

# DIVISIBILITY TRICKS

How can you tell if a number is a factor of another number—especially if you haven't already memorized it from the multiplication table? You can always divide and see whether you get a zero remainder.

For example, $13$ is a factor of $156$ because you can work out the division:

$$\begin{array}{r} 12 \\ 13\overline{)156} \\ -13 \\ \hline 26 \\ -26 \\ \hline 0 \end{array}$$

Fortunately, there are a number of tricks that you can use to check divisibility for many small divisors.

## Trick Table

| | |
|---|---|
| 1 | **Every number is divisible by 1.** |
| 2 | **A number is divisible by 2 if its last digit is 0, 2, 4, 6, or 8.** If a number is divisible by 2 it is called **even**. Otherwise, it's **odd**. Even numbers end in even digits: 0, 2, 4, 6, 8. Odd numbers end in odd digits: 1, 3, 5, 7, 9. |
| 3 | **A number is divisible by 3 if the sum of its digits is divisible by 3.** So 474 is divisible by 3 because $4 + 7 + 4 = 15$, which is divisible by 3. But 895 is not divisible by 3 because $8 + 9 + 5 = 22$, which is not divisible by 3. |
| 4 | **A number is divisible by 4 if its last two digits, taken together as a two-digit number, are divisible by 4.** So 456 is divisible by 4 because 56 is. But 367 isn't divisible by 4 because 67 isn't. |
| 5 | **A number is divisible by 5 if it ends in a 5 or a 0.** |
| 6 | **A number is divisible by 6 if it's divisible by both 2 and 3.** So 234 is divisible by 6 because it's even and its digits add up to 9. |
| 9 | **A number is divisible by 9 if the sum of its digits is divisible by 9.** So 729 is divisible by 9 because its digits add up to 18. But 1024 is not divisible by 9: its digits add up to 7. |
| 10 | **A number is divisible by 10 if its last digit is a 0.** |

And, of course, every number is divisible by itself.

# PRIME NUMBERS

A **prime number** has no factors except for itself and $1$.

So $3$ is prime because its only factors are $1$ and $3$. But $24$ is not prime because it's also divisible by $2, 3, 4, 6, 8$, and $12$.

The number $1$ is not considered prime. Neither is $0$.

Every prime number except $2$ is odd. Is that odd? Not at all. Even numbers are divisible by $2$, so they can't be prime.

Here are all the primes less than $50$. It's kind of nice to know what they are by heart.

$$2, 3, 5, 7, 11, 13, 17, 19, 23, 29, 31, 37, 41, 43, 47$$

## CHECKING FOR PRIMES

Unfortunately, there are no fast tricks to check if a number is prime. You just have to check whether it has any factors, one by one. The good news is that you only have to check the *prime* factors. If it doesn't have any interesting prime factors, then it's prime.

So to determine whether a number is prime, start trying to divide it by primes—by $2$, by $3$, by $5$, by $7$—until the quotient you get is smaller than the next highest prime.

Is $91$ prime?

Start checking. It's not divisible by $2$ because it's not even. It's not divisible by $3$ because its digits add up to $10$. It's not divisible by $5$ because it ends in a $1$. It *is* divisible by $7$, however: $91 \div 7 = 13$.

So $91$ is not prime: it factors as $7 \times 13$.

Is $61$ prime?

Again, start checking. It's not divisible by $2$, $3$, or $5$. It's not divisible by $7$. Stop here: $61$ divided by $7$ gives something between $8$ and $9$—and the next prime number is $11$, which is too high.

So $61$ is prime.

## COMPOSITE NUMBERS

Numbers that aren't prime (and aren't $1$ or $0$) are called **composite**. A composite number can be factored in an interesting way, not just $1$ times itself.

So $24$ is composite because we can say that $24 = 3 \times 8$. But $3$ is not composite: the only way to factor $3$ is as $1 \times 3$, which is not interesting.

### WHO CARES ABOUT PRIMES?

It turns out that every whole number in the world (except for $0$ and $1$) is either prime or a product of primes. And any number factors as a product of primes in one and only one way. So once you know that $294 = 2 \times 3 \times 7 \times 7$, you also know that $294$ isn't divisible by $5$ or $11$ or any prime except $2$, $3$, and $7$.

If you want to be romantic about it, primes are the indestructible building blocks that make up all other numbers.

## FACTOR TREES

A **factor tree** is a nice way to organize prime factor information when you're looking for all the prime factors of a number. Here's a factor tree for the number $20$.

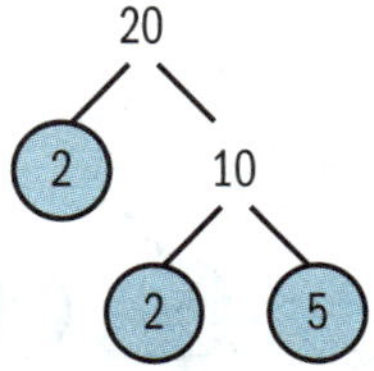


The final leaves of the tree are all prime factors of $20$. This factor tree tells you that $20 = 2 \times 2 \times 5$.

### MAKING A FACTOR TREE

Write down the number whose prime factors you want to find.

> If it's prime, then you're done. Circle it.
> If it's not, think of a way to factor it.

For each new "leaf" of the factor tree, do the same thing:

> If the leaf number is prime, then you're done with that leaf. Circle it.
> If it's not, find a factorization, draw branches, and write down the factors as leaves.

Rinse and repeat.

The number $20$ is not prime; it factors as $2 \times 10$, for example. Write down both $2$ and $10$ as branches coming out of $20$.

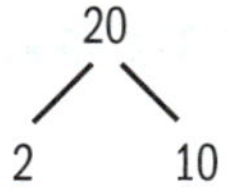


The number $2$ is prime, so we circle it. But $10$ factors as $5 \times 2$. Make new leaves.

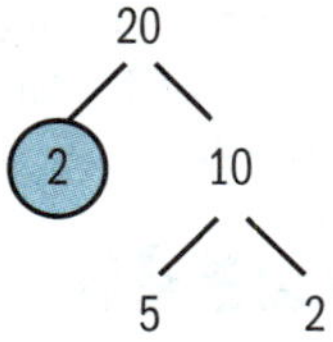


Both $2$ and $5$ are prime. Circle them. Done.

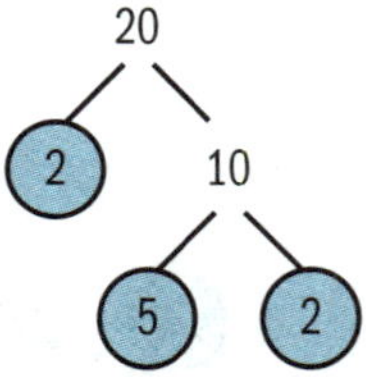


## IS A FACTOR TREE UNIQUE?

But what if you'd factored $20$ as $4 \times 5$ instead of as $2 \times 10$?

Not a problem. You may get a different factor tree—but still the same prime factors. Take a look:

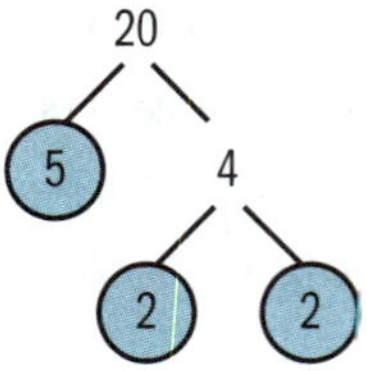


This tree tells us that $20 = 2 \times 2 \times 5$, just like the first tree.

Here's one last example: a factor tree for 108.

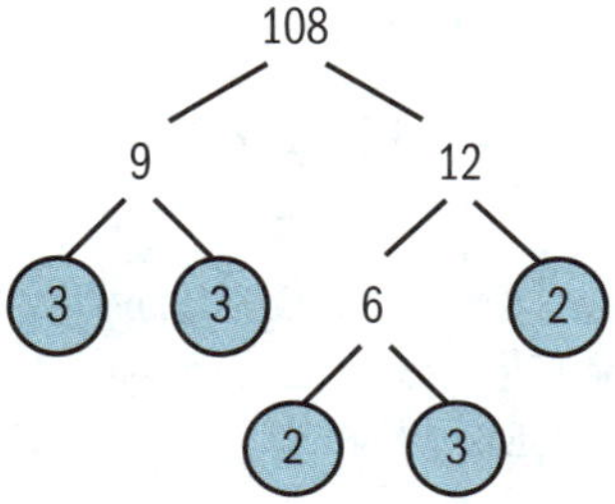


So $108 = 2 \times 2 \times 3 \times 3 \times 3$.

# GREATEST COMMON FACTOR

What is a **common factor** of two numbers? Easy—it's just what it says it is. It's a factor that they have in common.

For example, let's find the common factors of 42 and 30. First, we list all the factors of 42 and 30. Then we figure out which factors they share.

| **Factors of 42** | **Factors of 30** |
|---|---|
| 1, 2, 3, 6, 7, 14, 21, 42 | 1, 2, 3, 5, 6, 10, 15, 30 |

**Common factors:** 1, 2, 3, 6

The **greatest common factor (GCF)** of two numbers is another no-brainer: it's the greatest of their common factors. The GCF of 42 and 30 is 6.

You can also find common factors and the greatest common factor of more than two numbers. For example, the common factors of 6, 15, and 21 are 1 and 3. Their GCF is 3.

## FINDING THE GCF WITH PRIME FACTORIZATION

You can always find the GCF by listing all the common factors and finding the greatest one. But that's a lot of busywork. Often, it's faster to find the GCF using the prime factorization.

Let's go back to our previous example—42 and 30—and find their prime factorizations.

$$42 = 2 \times 3 \times 7$$
$$30 = 2 \times 3 \times 5$$

All you have to do now is pick out the *prime* factors that they have in common and multiply them together. The prime factors common to 42 and 30 are 2 and 3. So their GCF is $2 \times 3 = 6$. Easy!

Let's try a harder example. What's the GCF of 108 and 126?

We found the prime factorization of 108 in the section on factor trees above:

$$108 = 2 \times 2 \times 3 \times 3 \times 3.$$

Now we make a factor tree for 126:

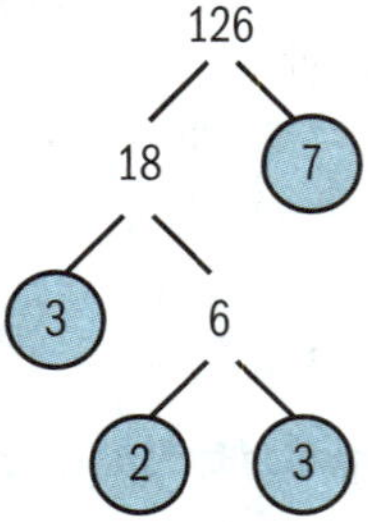


So $126 = 2 \times 3 \times 3 \times 7.$

Great! What are their common prime factors? Well, 2 and 3, evidently. But notice that 3 appears twice in 126 and at least twice in 108. So 108 and 126 share *two* prime factors of 3. When we multiply their prime factors to get the GCF, we have to include both of them. The GCF of 108 and 126 is $2 \times 3 \times 3 = 18$.

# LEAST COMMON MULTIPLE

Sometimes we need to know the **least common multiple (LCM)** of two numbers. The LCM is the least of the common multiples—the smallest number that's divisible by both numbers.

To find the LCM, factor both numbers into their prime factorization. Each prime factor that appears in either one must show up again in the LCM.

Huh? To explain, let's go back to our old friends $42$ and $30$.

$$42 = 2 \times 3 \times 7$$
$$30 = 2 \times 3 \times 5$$

The LCM of $42$ and $30$ will have to include all prime factors that appear in their two prime factorizations. So their LCM is $2 \times 3 \times 5 \times 7 = 210$.

Let's just check. The number $210$ should be divisible by both $42$ and $30$. And it is! Look: $210 = 42 \times 5$ and $210 = 30 \times 7$. Don't you love how it all works out?

Let's try a harder example: $24$ and $126$.

$$24 = 2 \times 2 \times 2 \times 3$$
$$126 = 2 \times 3 \times 3 \times 7$$

The LCM will have to include a $2$ and a $3$ and a $7$. But because $2$ appears three times (in the factorization of $24$) and $3$ appears two times (in $126$), we have to include those multiplicities when calculating the LCM.

So the LCM of $24$ and $126 = 2 \times 2 \times 2 \times 3 \times 3 \times 7 = 504$.

## ALIAS LCD

When working with fractions, you might have to look for what's called the **lowest common denominator (LCD)**. Don't worry—the LCD is the same thing as the LCM. It just has another name in a different context.

But more on denominators in the next chapter.

## CHECKING YOUR WORK: LCM AND GCF

One last check: it turns out that the product of two numbers is the same as the product of their GCF and their LCM:

$$(\text{first number}) \times (\text{second number}) = \text{GCF} \times \text{LCM}.$$

Take a look: the GCF of $42$ and $30$ is $6$; their LCM is $210$. Does $42 \times 30 = 6 \times 210$? You bet. Neat, eh?

# YOUR TURN

Solutions start on page 275.

1. List all the factors of $144$. How many are there?

2. Which of the following numbers are divisible by $2$? by $3$? by $4$? by $6$? Without dividing, can you guess which one is divisible by $12$?

   (a) $1{,}481{,}472$
   (b) $4{,}938{,}268$
   (c) $6{,}920{,}027$
   (d) $9{,}654{,}609$
   (e) $97{,}972{,}810$
   (f) $6{,}333{,}402$

3. For each number, determine whether it is prime. If it isn't, find its smallest prime factor.

   (a) $67$
   (b) $91$
   (c) $163$
   (d) $191$
   (e) $1001$

4. Make *three different* factor trees for $144$. What is its prime factorization?

5. Find the prime factorization of each number.

   (a) $66$
   (b) $113$
   (c) $175$
   (d) $936$

6. (a) List all the factors of $300$. List all the factors of $225$.
   (b) What are all the common factors of $300$ and $225$? What is their greatest common factor?
   (c) Find the prime factorizations of $300$ and $225$ and use them to find the prime factorization of their GCF.
   (d) Check that the GCF found in (b) matches the prime factorization in (c).

7. (a) Use the prime factorizations found in 6(c) to find the LCM of $300$ and $225$.
   (b) Find the product of the LCM from (a) and the GCF from 6(b).
   (c) Compare the product from (b) with the product $300 \times 225$.

8. Find the GCF and the LCM of each pair of numbers.
   (a) $5$ and $17$
   (b) $12$ and $60$
   (c) $35$ and $1001$

9. Kyle found that the GCF of two numbers is $21$ and that their LCM is $588$. One of the numbers is $147$. What is the other number?

10. The number $10!$ (ten factorial) is the product of

$$1 \times 2 \times 3 \times 4 \times 5 \times 6 \times 7 \times 8 \times 9 \times 10.$$

   (a) How many factors of $2$ are there in its prime factorization? How many factors of $3$? How many factors of $5$?
   (b) Find its prime factorization.
   (c) Extra-tricky bonus question: How many zeroes are at the end of $10!$ if you multiply it out? (Obviously you shouldn't multiply it out to answer this question.)

FACTORING

# 7

# CHAPTER 7
## FRACTIONS



### AGENT ORANGE AND THE BLACK MANTIS

After I finished my email to Rebecca, I tried to get some answers.

"So my dad didn't run off with you and leave my mom?" I asked my dad's beautiful blond secretary.

"Nope," she said. "I'm not a one-man kind of gal."

Yikes!

"I'm sorry we lied, honey," Mom said. "We didn't want to get you involved in our spy games."

"But when Agent Orange got too close to you, we had to intervene," Dad explained.

"Agent who?" I asked.

"You know him as Kyle Thomas," Dad's beautiful blond secretary explained.

"Kyle's a spy?" I exclaimed.

"It's not true!" he cried from his cage. I guess he had come to while I was typing away about primes.

"Shut up, Orange!" my mom yelled. "Or it's Mongolian belly-button torture for you!"

"Mom!" I said. But her icky threat seemed to do the trick. Kyle slinked back nearer to the wet wall, where Jake was starting to wake up.

"Sorry, dear," Mom apologized. "It's a very effective and sometimes lethal torture technique, but you don't need to worry your pretty little head about it."

It was all just too insane to believe. To clear my head, I took a walk around the empty caverns in the drippy cave, trying not to bonk into the stalactites hanging from the ceiling. When I calmed down a little bit, I returned to the TV room, where Dad and his secretary were following the red blip on the many screens.

"Who are you guys tracking there?"

"The Black Mantis," Dad said.

"Who's that?"

"Only the most dangerous spy alive today," his secretary said.

"You know her as Rebecca Romaine," Mom said.

"Shut up!" I exclaimed. "Rebecca's not a spy!" Then I started thinking. "Wait . . . BM—that's the monogram on her bag! BM for Black Mantis. And yesterday Kyle accidentally called her Mantis! But she can barely add and subtract . . ."

"Math skills are not a prerequisite for spying," Dad said. "Ruthless cunning, high pain threshold, sound belly-button-twisting technique—that's what makes a good spy."

"Now that you know about us and Rebecca, we are going to need your help, dear," Mom said, kneeling down before me. "We need you to lure her here to the drippy cave."

"Tell her you want to talk about fractions," Dad said. "It could save the world."

I took a deep breath and sent Rebecca—I mean, the Black Mantis—another email.

* * * * *

FROM: Shanon Fletcher [shanon4@britrockmail.com]
TO: Rebecca Romaine [rr009@blackmantis.net]
SUBJECT: Fractions (You wanna come over to talk about this stuff later?)

# FRACTIONS

A **fraction** describes a part of a whole.

I'd bet you've already seen fractions compared to pizza slices or pieces of cake—but it's really a pretty good analogy. So I'm going to have to be unoriginal like that.

If a whole cake is divided into eight slices, then each slice represents one-eighth of the cake.

This is written as $\frac{1}{8}$.

The top part of $\frac{1}{8}$ is called the **numerator**. The bottom part is called the **denominator**. (Memory trick: The *d* in *denominator* stands for "down.")

The denominator tells you into how many pieces the whole is cut. The cake has eight slices, so the denominator is $8$. The numerator tells you how many of the pieces we're dealing with right now. We were talking about one slice, so the numerator is $1$.

Three slices of cake is three-eighths, or $\frac{3}{8}$, of the whole cake.

## TALKING AND WRITING

I just want to go over how to name some easy fractions. This is, um, a piece of cake.

| | | | |
|---|---|---|---|
| $\frac{1}{2}$ | one-half | $\frac{5}{6}$ | five-sixths |
| $\frac{1}{3}$ | one-third | $\frac{4}{7}$ | four-sevenths |

FRACTIONS

$\frac{2}{3}$ two-thirds

$\frac{1}{4}$ one-fourth, a quarter

$\frac{3}{4}$ three-quarters, three-fourths

$\frac{1}{5}$ one-fifth

$\frac{7}{8}$ seven-eighths

$\frac{2}{9}$ two-ninths

$\frac{1}{10}$ one-tenth

$\frac{12}{17}$ twelve-seventeenths

The top and bottom of a fraction can be separated by either a horizontal line or a slanted line. I always worry about confusing $^1\!/_{27}$ with $12$ followed by $7$, so I prefer writing $\frac{1}{27}$ to be safe.

# EQUIVALENT FRACTIONS & LOWEST TERMS

Four slices of an eight-slice cake is $\frac{4}{8}$ of the whole. But four slices is also half the cake, which is written as $\frac{1}{2}$. So what's the right way of expressing so much yummy goodness? Fortunately, both are right, and neither is more right. These two fractions are **equivalent**. They represent the same amount.

$$\frac{4}{8} = \frac{1}{2}$$

One whole cake is represented by $1$. But it's also eight slices, which is $\frac{8}{8}$. So $1 = \frac{8}{8}$. Two perfectly good ways of naming the same chunk of cake.

## CHECKING EQUIVALENCE

To check if two fractions are equivalent, **cross-multiply**: multiply the numerator of one by the denominator of the other. If you get the same answer, the two fractions are equivalent.

So you can check that $\frac{4}{8} = \frac{1}{2}$ because $4 \times 2 = 1 \times 8$.

## REDUCING FRACTIONS TO LOWEST TERMS



Often, it's easiest to think of a fraction in its **lowest terms**. A fraction is in lowest terms if its numerator and denominator have no factors in common. The fraction $\frac{1}{2}$ is in lowest terms, but $\frac{4}{8}$ is not: both $4$ and $8$ are divisible by $2$.

A fraction not in lowest terms can be **simplified** or **reduced**. (Teachers frequently want answers to fraction problems in simplified form, so it's a good thing to know how to do.) To reduce a fraction, divide the numerator and the denominator by the same number. You will get an equivalent fraction with a smaller denominator. Keep reducing until the numerator and the denominator have no more common factors.

Let's try reducing $\frac{4}{8}$ by dividing top and bottom by $2$:

$$\frac{4}{8} = \frac{4 \div 2}{8 \div 2} = \frac{2}{4}.$$

Is $\frac{2}{4}$ in lowest terms? No, because both $2$ and $4$ are divisible by $2$. Reduce again:

$$\frac{2}{4} = \frac{2 \div 2}{4 \div 2} = \frac{1}{2}.$$

Now, $\frac{1}{2}$ is in lowest terms, and we're done.

Here's another example: reduce $\frac{15}{72}$ if it's not in lowest terms.

Do $15$ and $72$ have any common factors? Yes, both are divisible by $3$. Divide both sides to get

$$\frac{15}{72} = \frac{15 \div 3}{72 \div 3} = \frac{5}{24}.$$

Is $\frac{5}{24}$ in lowest terms? Yes: 5 and 24 don't have any common factors. So we're done.

### LOWEST TERMS USING GCF

You can also reduce a fraction all the way to lowest terms in one swoop. Divide both the numerator and the denominator by their GCF (greatest common factor—see Chapter 6 to review). This can cut down on some busywork, although it does mean that you need to find a GCF before you start reducing.

Let's try one: reduce $\frac{24}{64}$.

What's the GCF of $24$ and $64$? If you go through making the two factor trees, you'll find that $24 = 2 \times 2 \times 2 \times 3$ and $64 = 2 \times 2 \times 2 \times 2 \times 2 \times 2$. So the GCF (count those $2$s carefully!) is $2 \times 2 \times 2 = 8$.

Now reduce:

$$\frac{24}{62} = \frac{24 \div 8}{64 \div 8} = \frac{3}{8}.$$

And $\frac{3}{8}$ is in lowest terms! Fabulous.

## IMPROPER FRACTIONS & MIXED NUMBERS

### IMPROPER FRACTIONS

In a **proper fraction**, the numerator is less than the denominator. All the fractions that we've talked about so far have been proper fractions. They name quantities less than $1$.

If the denominator is greater than or equal to the numerator, the fraction is **improper**. It's just a technical term; it doesn't mean that the fraction is bad.

An improper fraction names a quantity more than one. For example, $\frac{9}{8}$ of an eight-slice cake is nine slices—so one whole cake and an additional slice:

Any whole number can be expressed as an improper fraction. For example, $3 = \frac{3}{1}$, and $7 = \frac{7}{1}$.

## MIXED NUMBERS

Another way to express quantities more than 1 is with a **mixed number.** A whole cake plus an additional slice is $1\frac{1}{8}$ of a cake. This is read as "one and one-eighth." Makes sense.

The mixed number $1\frac{1}{8}$ and the improper fraction $\frac{9}{8}$ are just two different names for the same thing. They represent the same quantity: $1\frac{1}{8} = \frac{9}{8}$.

## CONVERTING MIXED NUMBERS TO IMPROPER FRACTIONS

Improper fractions are easier to multiply and divide than mixed numbers, so we sometimes need to convert mixed numbers into improper-fraction form.

To make the conversion, multiply the whole number part by the denominator and add that to the numerator of the fractional part. This sum becomes the new numerator. The denominator stays the same.

$$\text{whole}\,\frac{\text{numerator}}{\text{denominator}} = \frac{(\text{whole} \times \text{denominator}) + \text{numerator}}{\text{denominator}}$$

Let's work out an example: convert $5\frac{3}{4}$ to an improper fraction.

$$5\frac{3}{4} = \frac{(5 \times 4) + 3}{4} = \frac{23}{4}$$

### CONVERTING IMPROPER FRACTIONS TO MIXED NUMBERS

It's easier to understand how big your quantity is from a mixed number than from an improper fraction. So after getting an answer in improper-fraction form, we sometimes convert back to a mixed number.

To make the conversion, divide the numerator by the denominator, with remainder. The quotient becomes the whole number part, and the remainder becomes the new numerator. The denominator stays the same.

$$\frac{\text{numerator}}{\text{denominator}} = \text{quotient}\ \frac{\text{remainder}}{\text{denominator}}$$

Let's convert $\frac{23}{4}$ back to an improper fraction and see what happens.

Do the division: $23 \div 4 = 5$, remainder $3$. So the whole number part is $5$, and the new numerator is $3$. As we hoped, we got $5\frac{3}{4}$.

Another example: express $\frac{13}{3}$ as a mixed number.

First do the division: $13 \div 3 = 4$, remainder $1$. So $\frac{13}{3} = 4\frac{1}{3}$. To check, convert back: $4 \times 3 + 1 = 13$, so we're good.

## COMPARING FRACTIONS

### SAME DENOMINATOR

Which is bigger, $\frac{1}{4}$ or $\frac{3}{4}$?

You can talk this one through: three-quarters must be larger than one-quarter.

Similarly, whenever you compare two fractions with the same denominator, all you have to do is compare their numerators. The fraction with the larger numerator is always larger.

### SAME NUMERATOR

Which is bigger, $\frac{1}{3}$ or $\frac{1}{4}$?

This one's trickier, but you can also think it through. One-third is one of three equal slices of cake. One-fourth is one of four equal slices of cake. Since there are fewer slices that the total cake is divided into, the third must be larger than the fourth.

$$\frac{1}{3} > \frac{1}{4}$$

Similarly, we can say that $\frac{1}{2}$ is larger than $\frac{1}{3}$, and that $\frac{1}{5}$ is larger than $\frac{1}{67}$.

## DIFFERENT DENOMINATORS

You might be able to estimate that $\frac{7}{8}$ is bigger than $\frac{1}{13}$ just by thinking about cake: $\frac{7}{8}$ is almost a full cake, whereas $\frac{1}{13}$ is smaller even than a regulation one-eighth cake slice. But what if you have to compare fractions such as $\frac{3}{5}$ and $\frac{4}{7}$? Both are close to a half—but it's tricky to tell how close.

We already talked about using cross-multiplication to check whether two fractions are equivalent. (Multiply the numerator of the first fraction by the denominator of the second and vice versa, and then compare the two products. If the cross-multiplication products are equal, then the fractions are equivalent.)

The nice thing is that you can also use cross-multiplication to compare two fractions. Compute the two cross-multiplication products and compare them. The fraction that contributed its numerator to the bigger product is the bigger fraction.

So to compare $\frac{3}{5}$ and $\frac{4}{7}$, do the cross-multiplication:

$3 \times 7$ $\quad$ $4 \times 5$

$3 \times 7 = 21$, which is greater than $4 \times 5 = 20$

So $\frac{3}{5}$ is greater than $\frac{4}{7}$.

Last example. What's bigger, $\frac{7}{8}$ or $\frac{9}{11}$?

Cross-multiply: $7 \times 11 = 77$ and $9 \times 8 = 72$. The cross-multiplication product $77$ is greater than $72$, so $\frac{7}{8}$ is greater than $\frac{9}{11}$.

# YOUR TURN

Solutions start on page 280.

1. For each fraction, determine whether it is in lowest terms. If it isn't, reduce it.

   (a) $\frac{3}{9}$

   (b) $\frac{2}{17}$

   (c) $\frac{16}{72}$

   (d) $\frac{45}{75}$

   (e) $\frac{30}{163}$

   (f) $\frac{38}{361}$

2. Which of these fractions are equivalent to $\frac{6}{14}$?

   (a) $\frac{4}{7}$

   (b) $\frac{9}{21}$

   (c) $\frac{33}{77}$

   (d) $\frac{12}{28}$

   (e) $\frac{222}{504}$

   (f) $\frac{138}{322}$

3. Convert each improper fraction to a mixed number.

(a) $\frac{6}{4}$

(b) $\frac{21}{9}$

(c) $\frac{144}{18}$

(d) $\frac{22}{7}$

(e) $\frac{40}{11}$



4. Convert each mixed number to an improper fraction.

(a) $1\frac{2}{3}$

(b) $1\frac{3}{5}$

(c) $2\frac{1}{5}$

(d) $3\frac{1}{4}$

(e) $7\frac{12}{17}$

5. Put each set of fractions in order, from least to greatest.

(a) $\frac{9}{11}, \frac{6}{11}, \frac{7}{11}, \frac{1}{11}, \frac{2}{11}, \frac{15}{11}, \frac{11}{11}$

(b) $\frac{1}{3}, \frac{1}{7}, \frac{1}{2}, \frac{1}{11}, \frac{1}{4}, 1, 0$

(c) $\frac{11}{12}, \frac{4}{5}, \frac{1}{2}, \frac{10}{11}, \frac{8}{9}, 1$

(d) $\frac{1}{2}, \frac{1}{3}, \frac{2}{3}, \frac{2}{5}, \frac{3}{5}, \frac{2}{7}, \frac{3}{7}, \frac{4}{7}, \frac{5}{7}$

6. For each fraction, say whether it is less than or greater than $\frac{5}{8}$.

(a) $\frac{1}{2}$

(b) $\frac{2}{3}$

(c) $\frac{3}{5}$

(d) $\frac{4}{9}$

(e) $\frac{7}{11}$

(f) $\frac{8}{13}$

7. At Johnny London's Friday night show a couple of months ago, $\frac{2}{5}$ of the people in the audience were ardent fans, like me, and $\frac{1}{3}$ of the people had never heard of them. Which was the bigger group?

8. Once, before all this madness started, Rebecca and I were eating sweet-potato fries at the Finer Diner. I ate $\frac{1}{2}$ of the fries in the basket; she ate $\frac{2}{4}$ of the fries in the basket. How much was left?

# 8

# CHAPTER 8
# ADDING AND SUBTRACTING FRACTIONS

## CONFRONTING AGENT ORANGE

I sent the email, and we all waited for Rebecca to reply. Dad's beautiful blond secretary did her nails. (Wow, she really is beautiful!) Mom went back to crocheting a frog onto a throw pillow for our family room sofa. Dad disappeared into the outhouse with a copy of *Spy and Garden*. I went over to talk to my new ex-boyfriend, Agent Orange.

"Why didn't you tell me you were a spy trying to take over the world?" I asked. "You know, it's not that you're a spy trying to take over the world that bothers me. It's that you *lied* to me."

"I didn't think you'd believe me, mate," he said. "I'm so ashamed. And I really liked you, too. In fact, I've never met a bird like you."

"I liked you too, Kyle!—though I'm not a bird; I'm a girl," I said. "And I really thought that Johnny London was awesome."

"I know. We rock!" he said. "The band started out as our cover. We were going to use it to infiltrate the United States media oligopoly and plant executives loyal to our cause in key positions at major news and entertainment corporations. . . . How could I know I was going to turn into a great singer-slash-songwriter-slash-guitar-player? If I had guessed that I have this amazing talent, I would have never gotten involved in the world domination game! Never!"

Jake, who had been lying in a heap off to the side, tried to raise his head. "Yeah, we really *are* good, aren't— Oww! My head starts killing me when I talk."

"Then don't," I snapped. "Anyway, so Kyle, were you really going to let me listen to the recording of you guys playing at the Aragorn, or was that just another spy lie?"

"It was true!" Kyle said. "But I can't lie anymore. After I played you our show, I was going to kidnap you and hold you for ransom."

"I am so disappointed in you," I said.

Mom called over from the computer, "Honey, you got an email back from the Black Mantis!"

I went over to read it.

"Now she wants help adding and subtracting fractions!" I exclaimed. "What should I do?"

"Help her!" Dad said. "We can't let her find out we're on to her!"

## ADDING FRACTIONS: SAME DENOMINATOR

We're ready to start working with fractions. Let's start with addition.

What is $\frac{2}{7} + \frac{3}{7}$?

All sevenths are the same, so two-sevenths plus three-sevenths should give five-sevenths, right? Indeed.

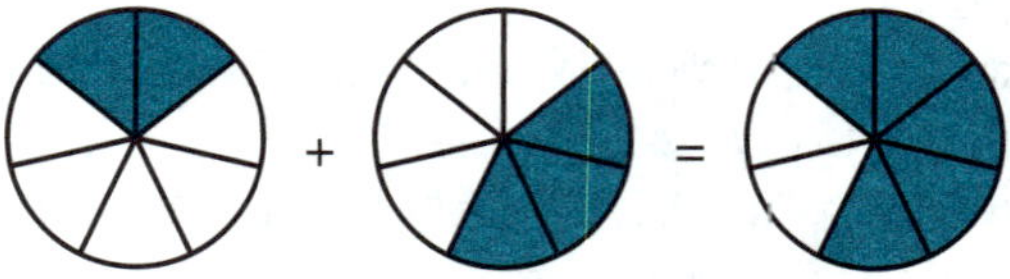

$$\frac{2}{7} \quad + \quad \frac{3}{7} \quad = \quad \frac{5}{7}$$

When the denominators are the same, add fractions just by adding their numerators:

$$\frac{2}{7} + \frac{3}{7} = \frac{5}{7}.$$

Easy enough.

## ADDING AND REDUCING

Another example: What's $\frac{4}{9} + \frac{2}{9}$?

Since the denominators are the same, we just add the numerators:

$$\frac{4}{9} + \frac{2}{9} = \frac{6}{9}.$$

Excellent: $\frac{6}{9}$ *is* the right answer. But it's not in lowest terms: both $6$ and $9$ are divisible by $3$. So we can reduce:

$$\frac{6}{9} = \frac{6 \div 3}{9 \div 3} = \frac{2}{3}.$$

So $\frac{4}{9} + \frac{2}{9} = \frac{2}{3}$.

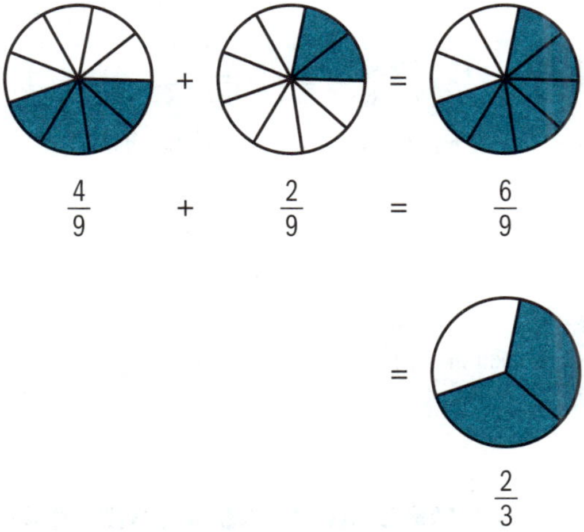


# ADDING FRACTIONS: DIFFERENT DENOMINATORS

Let's jump right in: what's $\frac{2}{5} + \frac{1}{3}$?

The answer is most emphatically *not* $\frac{3}{8}$.

To add two fractions with different denominators, we first need to convert the fractions to equivalent fractions with the same denominator. Luckily, we've already gone over all the skills you need to know to pull these kinds of tricks.

What same denominator should we use? To be able to make the conversion, the new denominator has to be a multiple of both of the original denominators. What number is a multiple of both $3$ and $5$? Certainly their product will work. Their product is $15$. Let's use that.

We convert each fraction separately.

$\frac{2}{5}$: The denominator $5$ has to be multiplied by $3$ to get $15$. To ensure that the new fraction is still equivalent to $\frac{2}{5}$, we multiply both top and bottom by $3$:

$$\frac{2}{5} = \frac{2 \times 3}{5 \times 3} = \frac{6}{15}.$$

$\frac{1}{3}$: The denominator $3$ has to be multiplied by $5$ to get $15$. So we multiply both top and bottom by $5$ to get an equivalent fraction:

$$\frac{1}{3} = \frac{1 \times 5}{3 \times 5} = \frac{5}{15}.$$

We can now rewrite the problem and do the addition:

$$\frac{2}{5} + \frac{1}{3} = \frac{6}{15} + \frac{5}{15} = \frac{11}{15}.$$

We sort of plodded through that one, so let's go over how to do this stuff again, more carefully.

## METHOD 1: MORE BUSYWORK, LESS THINKING

When adding two unlike fractions,

1. Convert both to equivalent fractions with a **common denominator**. The simplest common denominator is the product of the two original denominators.

2. Add.
3. Reduce and/or convert the sum to a mixed number.

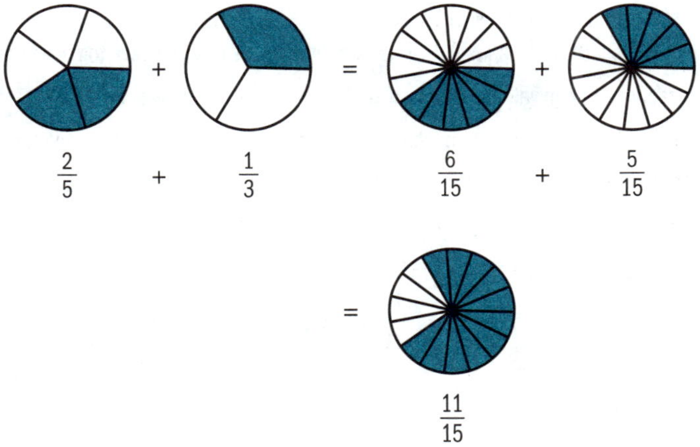


The first two steps can be expressed like this:

$$\frac{\text{numerator 1}}{\text{denominator 1}} + \frac{\text{numerator 2}}{\text{denominator 2}} =$$

$$= \frac{(\text{numerator 1}) \times (\text{denominator 2})}{(\text{denominator 1}) \times (\text{denominator 2})} + \frac{(\text{numerator 2}) \times (\text{denominator 1})}{(\text{denominator 1}) \times (\text{denominator 2})}$$

Take a look: What you do to the numerators is reminiscent of the cross-multiplying trick we used for comparing fractions (see page 106)—except this time, you add the two cross-multiplication products together.

The final sum can be expressed like this:

$$\frac{(\text{numerator 1}) \times (\text{denominator 2}) + (\text{numerator 2}) \times (\text{denominator 1})}{(\text{denominator 1}) \times (\text{denominator 2})}.$$

Let's see how this works with the $\frac{2}{5} + \frac{1}{3}$ example that we just did. Plug in $2$ for (numerator $1$), $5$ for (denominator $1$), $1$ for (numerator $2$), and $3$ for (denominator $2$):

$$\frac{2}{5} + \frac{1}{3} = \frac{2 \times 3 + 1 \times 5}{5 \times 3} = \frac{11}{15}.$$

We got the same answer—it looks like this method works.

## METHOD 2: MORE THINKING, LESS BUSYWORK

The method that we just went over (Method 1) will always, always work. But sometimes it's *so* inefficient.

Suppose you're adding $\frac{1}{8} + \frac{1}{16}$. Method 1 says that you have to convert both fractions to equivalent fractions with the common denominator $8 \times 16 = 128$. That's huge. That's three *digits*! You have to lug this monster of a denominator around as you add, and then—and then!—you have to reduce back down because the answer will be in very not-lowest terms. Just watch:

**Add:**

$$\frac{1}{8} + \frac{1}{16} = \frac{1 \times 16 + 1 \times 8}{8 \times 16} = \frac{24}{128}$$

**Reduce:**

$$\frac{24}{128} = \frac{24 \div 8}{128 \div 8} = \frac{3}{16}$$

So first we had to multiply 8 by 16, then we divide the answer by 8 again. Exhausting and wasteful!

If you're lazy like me, you can save on some work by choosing a better common denominator.

As I've already mentioned, any common denominator has to be a common multiple of both of the original denominators. The smallest one that we can use is (duh!) the least common multiple (LCM). That's why the LCM is frequently called the **lowest common denominator**, or **LCD**. Fortunately, you already know how to find it (take a look at Chapter 6 if you forget).

The LCD of 8 and 16 is 16. So when adding $\frac{1}{8} + \frac{1}{16}$, you don't even have to convert $\frac{1}{16}$: it already has the right denominator. The fraction $\frac{1}{8}$ *does* need to be converted. Since 8 must be multiplied by 2 to get 16, multiply top and bottom by 2 to make the conversion:

$$\frac{1}{8} = \frac{1 \times 2}{8 \times 2} = \frac{2}{16}.$$

Adding is now simple:

$$\frac{1}{8}+\frac{1}{16}=\frac{2}{16}+\frac{1}{16}=\frac{3}{16}.$$

Much better. And no messy reduction necessary.

To recap, when adding fractions with the LCD method . . .

1. Find the LCD of the denominators.
2. Convert each fraction to an equivalent fraction with the LCD denominator.
3. Add.
4. Reduce if necessary.

## COMPARING THE TWO METHODS

So which method is better? Neither. Use whichever one is less painful for you. It's nice to know the LCD method–it's considered more elegant, if you care about such things. But either method is fine, and either will give you the right answer.

For the sake of comparison, let's race through an example both ways.

Compute $\frac{3}{8}+\frac{5}{6}$.

**Method 1:** With denominator $48 = 8 \times 6$

- **Convert:**

$$\frac{3}{8}=\frac{3\times 6}{8\times 6}=\frac{18}{48} \quad \text{and} \quad \frac{5}{6}=\frac{5\times 8}{6\times 8}=\frac{40}{48}.$$

- **Add:**

$$\frac{3}{8}+\frac{5}{6}=\frac{18}{48}+\frac{40}{48}=\frac{58}{48}.$$

- **Reduce:**

$$\frac{58}{48}=\frac{58\div 2}{48\div 2}=\frac{29}{24}, \text{ which is the same thing as } 1\frac{5}{24}.$$

**Method 2:** With LCD of 8 and 6

- **Find the LCD:** LCD of 8 and 6 is 24.

ADDING & SUBTRACTING

- **Convert.**

  $\frac{3}{8}$ : multiply top and bottom by $3$ to get a denominator of $24$

$$\frac{3}{8} = \frac{3 \times 3}{8 \times 3} = \frac{9}{24}$$

  $\frac{5}{6}$: multiply top and bottom by $4$ to get a denominator of $24$

$$\frac{5}{6} = \frac{5 \times 4}{6 \times 4} = \frac{20}{24}$$

- **Add:**

$$\frac{3}{8} + \frac{5}{6} = \frac{9}{24} + \frac{20}{24} = \frac{29}{24}$$

- **Reduce:**

  No reduction necessary, but $\frac{29}{24}$ is the same thing as $1\frac{5}{24}$.

## ADDING THREE OR MORE FRACTIONS

Adding many fractions works the same way as adding two.

What's $\frac{3}{5} + \frac{3}{8} + \frac{1}{10}$?

**Find the LCD:** The LCD of $5$, $8$, and $10$ is $40$.

**Convert and add:**

$\frac{3}{5}$: multiply top and bottom by $8$

$\frac{3}{8}$: multiply top and bottom by $5$

$\frac{1}{10}$: multiply top and bottom by $4$

So

$$\frac{3}{5}+\frac{3}{8}+\frac{1}{10}=\frac{3\times 8+3\times 5+1\times 4}{40}=\frac{24+15+4}{40}=\frac{43}{40}.$$

**Simplify:** $\frac{43}{40}$ is the same thing as $1\frac{3}{40}$.

# SUBTRACTING FRACTIONS

There's nothing new to learn for subtracting fractions. If the denominators are the same, subtract the numerators directly. Otherwise, convert to equivalent fractions with a common denominator before subtracting.

As with addition, the common denominator can be any common multiple of the denominators. And again the least thought-intensive option is the product of the two original denominators, and again using the LCD can save you some busywork.

What's $\frac{3}{4}-\frac{7}{11}$?

The LCD of $4$ and $11$ is $44$. Convert and subtract:

$$\frac{3}{4}-\frac{7}{11}=\frac{3\times 11-7\times 4}{4\times 11}=\frac{5}{44}.$$

The difference $\frac{5}{44}$ is already in lowest terms, so no need to reduce.

Compute $\frac{3}{2}-\frac{1}{6}$.

The LCD of $2$ and $6$ is $6$. Convert and subtract:

$$\frac{3}{2}-\frac{1}{6}=\frac{3\times 3-1}{6}=\frac{8}{6}.$$

The difference $\frac{8}{6}$ reduces to $\frac{4}{3}$, which is the same thing as $1\frac{1}{3}$.

# MIXED NUMBERS

When working with mixed numbers, it's often easier to work with the whole-number parts and the fractional parts separately—especially if the numbers are big.

## ADDING

Add the whole-number parts and the fractional parts separately. If the sum of the fractional parts is an improper fraction, convert it to a mixed number and combine with the whole-number sum.

For example: $1\frac{3}{5} + 2\frac{1}{2} = (1 + 2) + \left(\frac{3}{5} + \frac{1}{2}\right)$.

The fraction sum works out to $\frac{3}{5} + \frac{1}{2} = \frac{3 \times 2 + 1 \times 5}{5 \times 2} = \frac{11}{10}$. But $\frac{11}{10}$ is an improper fraction, and converts to $1\frac{1}{10}$. So we have to add that on to the sum of the whole numbers:

$$1\frac{3}{5} + 2\frac{1}{2} = (1 + 2) + \left(\frac{3}{5} + \frac{1}{2}\right) = 3 + 1\frac{1}{10} = 4\frac{1}{10}.$$

## SUBTRACTING

If the fractional part of the first number is smaller than the fractional part of the second number, you have to borrow from the whole-number part of the first number. (This borrowing is similar to the regrouping that we have to do when subtracting many-digit numbers in a stack.)

Like so: $7\frac{1}{4} - 4\frac{1}{3} = ?$

Since $\frac{1}{4}$ is less than $\frac{1}{3}$, we have to think of $7\frac{1}{4}$ as $6 + 1\frac{1}{4}$ in order to do the problem. The mixed number $1\frac{1}{4}$ is the same thing as $\frac{5}{4}$, so

$$7\frac{1}{4} - 4\frac{1}{3} = \left(6 + \frac{5}{4}\right) - 4\frac{1}{3} = (6 - 4) + \left(\frac{5}{4} - \frac{1}{3}\right).$$

The difference of the whole numbers is $2$. The fractional subtraction works out to $\frac{5}{4} - \frac{1}{3} = \frac{15 - 4}{12} = \frac{11}{12}$, so the answer is $2\frac{11}{12}$.

## TIRED? JUST BLUDGEON THROUGH

If you're feeling too lazy to think about all this whole-number and fractional part stuff, just convert all the mixed numbers to improper fractions, do the computations, and then convert back. You may do some extra busywork, but you'll get the right answer–which, after all, is the goal of all of these shenanigans.

For example, the problem $7\frac{1}{4} - 4\frac{1}{3}$ can also be worked out by converting $7\frac{1}{4}$ to $\frac{29}{4}$ and $4\frac{1}{3}$ to $\frac{13}{4}$, and then subtracting:

$$\frac{29}{4} - \frac{13}{3} = \frac{29 \times 3 - 13 \times 4}{12} = \frac{35}{12}.$$

Finally, $\frac{35}{12}$ is an improper fraction that converts to $2\frac{11}{12}$. The computation may be less pleasant, but it's also less confusing. And the answer is good.

# YOUR TURN

Solutions start on page 283.

1. Simplify. Leave your answers in lowest terms.

   (a) $\frac{2}{11}+\frac{5}{11}-\frac{3}{11}$

   (b) $\frac{19}{18}-\frac{2}{18}-\frac{5}{18}$

   (c) $\frac{21}{24}-\left(\frac{19}{24}-\frac{18}{24}\right)$

2. Simplify. Leave your answer in lowest terms.

   (a) $\frac{1}{3}+\frac{1}{5}$

   (b) $\frac{2}{11}+\frac{4}{13}$

   (c) $\frac{9}{14}-\frac{2}{7}$

   (d) $\frac{5}{8}+\frac{7}{40}$

   (e) $\frac{1}{21}-\frac{1}{28}$

   (f) $\frac{5}{6}-\frac{1}{10}$

3. Simplify. Leave your answer in lowest terms.

   (a) $\frac{1}{3}+\frac{1}{4}+\frac{1}{5}$

   (b) $1-\left(\frac{1}{2}+\frac{1}{3}\right)$

   (c) $\frac{1}{77}-\frac{1}{143}+\frac{1}{91}$

   (d) $\frac{7}{12}+\frac{23}{54}-\frac{3}{16}$

4. Simplify. Leave your answer as a mixed number in lowest terms.

(a) $3\frac{1}{3} - 2\frac{1}{4}$

(b) $\frac{17}{6} + \frac{13}{8}$

(c) $4\frac{1}{9} - \frac{19}{6}$

5. Simplify. Leave your answer as an improper fraction in lowest terms.

(a) $\frac{11}{12} + \frac{9}{8}$

(b) $3\frac{1}{2} + 1\frac{2}{3}$

(c) $4\frac{7}{22} - \left(2\frac{1}{4} - \frac{40}{33}\right)$

# 9

# CHAPTER 9
## MULTIPLYING AND DIVIDING FRACTIONS

### EVIL BLACK MANTIS

I got another email from Rebecca.

"Uh-oh, the Black Mantis is asking about Kyle and Jake," I called to my parents. "What should I do?"

"I was afraid of this," Dad said. "We're going to have to release those two, or she's going to figure out that we're on to them. Tell her that they're with you . . . tell her that you're watching a Johnny London rehearsal. And you want to meet her at the Finer Diner in one hour."

"I'll inject Agents Orange and Taff with Memorase," Mom said, picking up some kind of gun-like contraption. "That should do the trick."

She took aim and pressed the trigger twice quickly.

Kyle and Jake grabbed their bottoms. "Ouch!" they yelled. Then they passed out.

Mom walked over to their cage and plucked out the darts that were sticking out of their butts. "Okay, honey, drive these two to the Finer Diner," Mom said. "By the time they get there, the sedative will have worn off, and they won't remember a thing."

"Just tell them they must have fallen asleep in the backseat, as–uh–as you were driving them back from their rehearsal," Dad said. "Trust me, they'll agree with whatever you say. That's a side effect of Memorase–extreme suggestibility."

"This plan really sucks," I said. "What if Rebecca doesn't believe me and tries to kill me? What if Kyle and Jake try to take me hostage? What if I get so nervous that I forget what reciprocals are? Rebecca will know for sure that something's up!"

Mom came over and kissed me on the forehead.

"I know it's a lot to take in at once, honey," she said. "But the fate of humankind depends on you pulling off this tutoring session."

Mom, Dad, and I loaded the comatose members of Johnny London into my 2001 Saturn. I drove them to the Finer Diner in silence.

* * * * *

"Where have you been?" Rebecca asked. "Don't you know I have a big math test tomorrow? It's on dividing fractions, which is, like, my worst nightmare!"

"Sorry, Becca," I said. "We were rocking out in rehearsal."

"I know you're lying," Rebecca said. Her sinister eyes fixed on mine.

I stopped breathing. My former best friend was about to kill me. She knew that I knew that she was a master spy called the Black Mantis. And I knew that she knew that I knew. And maybe she even knew that I knew that she knew that I knew. I was going to die before I had the chance to become a famous Brit-pop producer.

"No, she's not lying, mate," Kyle said to Rebecca. "We were practicing and asked her do some backup vocals. This bird can really sing! We must have fallen asleep in the backseat of her 2001 Saturn on the way here."

"Like he said," Jake yawned, nodding at Kyle. "Man, my head hurts."

Rebecca looked at each of us in turn. Her sinister eyes narrowed. I felt my skin turning blue.

"Okay," she said. "I guess I have to believe you." I started breathing again. "Now, come on, Shanon, I need help multiplying and dividing fractions.—Waitress, coffees all around!"

"And a side order of taffy," Jake whispered.

"No taffy!" yelled the Black Mantis. She truly *was* evil.



## MULTIPLYING FRACTIONS

Multiplying fractions is the easiest thing ever: Multiply across the numerator and across the denominator. Just watch!

$$\frac{2}{7} \times \frac{3}{5} = \frac{2 \times 3}{7 \times 5} = \frac{6}{35}$$

Here's another:

$$\frac{8}{3} \times \frac{2}{17} = \frac{8 \times 2}{3 \times 17} = \frac{16}{51}.$$

And another!

$$\frac{1}{2} \times \frac{1}{3} = \frac{1}{6}$$

I could do this all day.

Of course, sometimes you'll have to reduce the product: for example, $\frac{1}{2} \times \frac{4}{5} = \frac{4}{10}$, which reduces to $\frac{2}{5}$.

That's all you *need* to know. I'm going to go over some tricks for cutting down on some of the reducing work, but it's all optional. Feel free to skip the next two sections.

## TRICK #1: REDUCE FIRST

If the fractions you're multiplying together aren't in lowest terms to begin with, reducing the product will be that much more cumbersome. Take a look at this unruly beast:

$$\frac{16}{96} \times \frac{36}{54} = \frac{576}{5184}.$$

Who wants to deal with taming a monster like $\frac{576}{5184}$? I shudder just thinking about it.

But take a look at how lovely everything becomes if you reduce first.

$\frac{16}{96}$ reduces to $\frac{1}{6}$

$\frac{36}{54}$ reduces to $\frac{2}{3}$

And the multiplication is an absolute dream:

$$\frac{16}{96} \times \frac{36}{54} = \frac{1}{6} \times \frac{2}{3} = \frac{2}{18}.$$

You still have to reduce $\frac{2}{18}$ to $\frac{1}{9}$, but isn't it *ever* so much nicer?

By the way, who would have guessed that $\frac{576}{5184}$ is the same thing as $\frac{1}{9}$? Ugh. Fractions can be tricky like that.

## TRICK #2: CANCEL FACTORS

What's $\frac{1}{2}$ times $\frac{2}{3}$?

Easy enough:

$$\frac{1}{2} \times \frac{2}{3} = \frac{2}{6}, \text{ which reduces to } \frac{1}{3} \text{ since } \frac{2}{6} = \frac{2 \div 2}{6 \div 2}.$$

Let's trace through what happens to the $2$s: there's one in the denominator of $\frac{1}{2}$ and one in the numerator of $\frac{2}{3}$. The fractions multiply to $\frac{1 \times 2}{2 \times 3}$, and then the product is reduced—the $2$s effectively cancel each other out: $\frac{1 \times \cancel{2}}{\cancel{2} \times 3} = \frac{1}{3}$.

To cut down on work, we often **cancel** common factors from numerators and denominators of the fractions before carrying out the multiplication. It's written like this:

$$\frac{1}{\cancel{2}} \times \frac{\cancel{2}}{3} = \frac{1}{3}.$$

Another example: Here's a slightly more involved problem: what's $\frac{4}{9} \times \frac{15}{22}$?

The common factors aren't as obvious as in the previous example, but they're still there. For example, $9$ and $15$ have a common factor of $3$, which you can cancel out, leaving $3$ and $5$. We write the leftover factors near the original numbers, like this:

$$\frac{4}{\underset{3}{\cancel{9}}} \times \frac{\overset{5}{\cancel{15}}}{22}.$$

We can now multiply across to get $\frac{4 \times 5}{3 \times 22}$, which we can simplify and then reduce. Or we can notice that $4$ and $22$ also have a common factor, and cancel out the $2$, leaving $2$ and $11$:

$$\frac{\overset{2}{\cancel{4}}}{\underset{3}{\cancel{9}}} \times \frac{\overset{5}{\cancel{15}}}{\underset{11}{\cancel{22}}}$$

Finally, we multiply across and compute $\frac{2 \times 5}{3 \times 11}$ to get the answer:

$$\frac{\overset{2}{\cancel{4}}}{\underset{3}{\cancel{9}}} \times \frac{\overset{5}{\cancel{15}}}{\underset{11}{\cancel{22}}} = \frac{10}{33}$$

**A wee word of caution:** If you're canceling a factor of, say, $5$, from the numerator of one fraction, you'd better cancel a $5$ from the denominator of another. So if you're multiplying together $\frac{1}{4}$ and $\frac{5}{8}$, there's nothing to cancel: although $4$ and $8$ share a factor of $4$, they're both in the denominator. They have to stay.

**Bottom line? Good news!** If you're clever about canceling common factors before multiplying, you'll never have to reduce the finished product: it will already be in lowest terms.

## WHOLE NUMBERS

This bit is dead easy!

What's $4 \times \frac{2}{3}$?

The whole number $4$ secretly belongs in the numerator (remember, it's the same thing as $\frac{4}{1}$), so

$$4 \times \frac{2}{3} = \frac{4}{1} \times \frac{2}{3} = \frac{4 \times 2}{1 \times 3} = \frac{8}{3}.$$

Told you it was easy.

## IMPROPER FRACTIONS

This bit is even easier! Multiply improper fractions exactly the same way as any other fraction.

Take a gander at $\frac{9}{8} \times \frac{28}{5}$. Just for practice, let's cancel out the common factor of $4$ from $8$ and $28$.

$$\frac{9}{\underset{2}{\cancel{8}}} \times \frac{\overset{7}{\cancel{28}}}{5} = \frac{63}{10}, \text{ which is the same thing as } 6\frac{3}{10}.$$

## MANY FRACTIONS IN ONE GO

Another no-brainer. You can multiply three or more fractions together in one swoop: the numerator of the product is the product of the numerators; the denominator is the product of the denominators:

$$\frac{5}{9} \times \frac{1}{15} \times \frac{3}{2} = \frac{15}{270}.$$

You can either simplify $\frac{15}{270}$, or you can do some canceling ahead of time. Let's see what that's like.

Cancel the factor of $5$:

$$\frac{\cancel{5}}{9} \times \frac{1}{\underset{3}{\cancel{15}}} \times \frac{3}{2}.$$

Cancel out the factor of $3$. In the denominator, you can take a $3$ either from the $9$ or from what's left of the $15$. Either one is fine.

$$\frac{\cancel{5}}{9} \times \frac{1}{\underset{\cancel{3}}{\cancel{15}}} \times \frac{\cancel{3}}{2}$$

Multiply:

$$\frac{\cancel{5}}{9} \times \frac{1}{\underset{\cancel{3}}{\cancel{15}}} \times \frac{\cancel{3}}{2} = \frac{1}{18}.$$

No reducing necessary.

# DIVIDING FRACTIONS

Dividing fractions works the same way as multiplying—with one little twist: you have to flip the fraction you're dividing by. Read on!

## RECIPROCALS

The **reciprocal** of a fraction is the fraction flipped over, top to bottom. Like so:

$$\frac{2}{3} \text{ flips over to become } \frac{3}{2}.$$

We say that $\frac{3}{2}$ is the reciprocal of $\frac{2}{3}$. Contrariwise, $\frac{2}{3}$ is the reciprocal of $\frac{3}{2}$.

The reciprocal of a number is also sometimes called its **inverse**.

### Whole number reciprocals

Sometimes the reciprocal of a fraction is a whole number.

This always happens when the numerator of the fraction is 1: for example, the reciprocal of $\frac{1}{5}$ is $\frac{5}{1}$, which is the same thing as 5. But it can also happen in other cases if the fraction is not in lowest terms. So the reciprocal of $\frac{2}{18}$ is $\frac{18}{2}$, which is the same thing as 9.

### Reciprocal of a whole number

The reciprocal of a whole number is the fraction $\frac{1}{\text{the whole number}}$. So the reciprocal of $10$ is $\frac{1}{10}$ and the reciprocal of $6$ is $\frac{1}{6}$. This makes perfect sense if you remember that 6 is the same thing as $\frac{6}{1}$.

### Special reciprocals

The number $0$ has no reciprocal. Since you can't divide by zero, it can *never* appear as the denominator of a fraction.

The reciprocal of $1$ is $1$. (Think of it as $\frac{1}{1}$.) The number $1$ is the only positive number that is its own reciprocal.

### Pairs of reciprocals

As a matter of fact, every number in the world (except $0$) is part of a reciprocal pair. So $\frac{3}{4}$ and $\frac{4}{3}$ are a reciprocal pair, as are $\frac{156}{2078}$ and $\frac{2078}{156}$, and $7$ and $\frac{1}{7}$. Ignoring the pair $1$ and $1$, one of the numbers in each reciprocal pair is a proper fraction (less than $1$); the other number is an improper fraction or a whole number (greater than $1$).

This pairing is why we call them *reciprocals*—the two numbers have a mutual (or "reciprocal") relationship.

## DIVIDING FRACTIONS

To divide by a fraction, multiply by its reciprocal.

So

$$\frac{2}{3} \div \frac{7}{5}$$

is the same thing as

$$\frac{2}{3} \times \frac{5}{7},$$

which is equal to $\frac{10}{21}$.

Since you already know everything about multiplying fractions and everything about reciprocals, you're basically good to go.

## CANCELING COMMON FACTORS

One thing to watch out for: make sure to do any canceling of common factors only *after* you flip. Otherwise you may get confused about where you're canceling from—the numerator or the denominator.

Let's look at a treacherous example:

$$\frac{5}{18} \div \frac{2}{15}.$$

It's tempting to cancel the $5$ from $5$ and $15$, isn't? But *don't do it*! Resist! When you flip the second fraction,

$$\frac{5}{18} \times \frac{15}{2},$$

the $15$ ends up in the numerator—on the same side of the bar as the $5$. And you can't cancel like that.

Did you notice that you *can* cancel a $3$ from $15$ and $18$, though? Go right ahead.

$$\frac{5}{\overset{}{\cancel{18}}_{6}} \times \frac{\overset{5}{\cancel{15}}}{2}$$

The final answer is $\frac{25}{12}$.

## WHY FLIPPING WORKS

If you don't really care, just skip this section. I won't mind.

Think about a problem such as $6 \div 3$. You're dividing $6$ into $3$ equal parts, and you know that each part is $2$. Let's think about what happens when you multiply by the reciprocal instead of dividing.

The problem

$$6 \div 3$$

becomes

$$6 \times \frac{1}{3},$$

which evaluates to $\frac{6}{3}$, or $2$.

The same thing works with fractions. For example, $\frac{3}{4} \div 3$ asks you what happens when you divide three-quarters into three equal parts. (Or, if $\frac{3}{4}$ is

3 parts, how big is one part?) Each part is $\frac{1}{4}$, which is exactly what you get when you multiply $\frac{3}{4} \times \frac{1}{3}$.

Something like $3 \div \frac{1}{2}$ is trickier. What does it mean to divide 3 into $\frac{1}{2}$ of an equal part? Essentially, the question can be rephrased: if 3 is $\frac{1}{2}$ of a part, what is a whole part? Well, if 3 is half, the whole thing is 6. And that's what you get when you do this flipping business:

$$3 \div \frac{1}{2} \text{ is the same thing as } 3 \times \frac{2}{1},$$

which multiplies out to 6.

# MIXED NUMBERS

When your multiplication or division involves mixed numbers, convert all the mixed numbers to improper fractions first. Then perform whatever computation you have to do. If the answer is an improper fraction, you can convert it back into a mixed number.

You already know how to do *all* of these things, so I'm going to do one example only. And it's more to practice division than anything else.

What is $9\frac{3}{5}$ divided by $4\frac{4}{9}$?

Let's dig right in.

**Convert:**

$$9\frac{3}{5} \text{ is the same thing as } \frac{9 \times 5 + 3}{5}, \text{ or } \frac{48}{5};$$

$$4\frac{4}{9} \text{ is the same thing as } \frac{4 \times 9 + 4}{5}, \text{ or } \frac{40}{9}.$$

**Flip:**

$$\frac{48}{5} \div \frac{40}{9} \text{ is the same thing as } \frac{48}{5} \times \frac{9}{40}.$$

**Multiply** (canceling out a factor of 8):

$$\frac{\overset{6}{\cancel{48}}}{5} \times \frac{9}{\underset{5}{\cancel{40}}} = \frac{54}{25}.$$

**Convert back:**

$$\frac{54}{25} \text{ is the same thing as } 2\frac{4}{25}.$$

Whew. So $\left(9\frac{3}{5}\right) \div \left(4\frac{4}{9}\right) = 2\frac{4}{25}$. Is this plausible? We were dividing a number between $9$ and $10$ by a number between $4$ and $5$. So we expect the answer to be close to $2$. Perfect.

MULTIPLYING & DIVIDING

# YOUR TURN

Solutions start on page 286.

1. Give the reciprocal of each number.

   (a) $\frac{4}{7}$

   (b) $\frac{2}{6}$

   (c) $0$

   (d) $1$

   (e) $1\frac{1}{5}$

   (f) $9$

2. Find each product. Leave your answers in lowest terms.

   (a) $\frac{1}{4} \times 3$

   (b) $\frac{2}{3} \times \frac{5}{7}$

   (c) $\frac{2}{3} \times \frac{2}{3}$

   (d) $\frac{1}{8} \times \frac{1}{8} \times \frac{1}{8}$

   (e) $\frac{1}{2} \times \frac{3}{4} \times \frac{5}{7}$

3. Simplify. Leave your answers in lowest terms.

(a) $\frac{3}{9} \times \frac{2}{5}$

(b) $\frac{9}{49} \times \frac{7}{3}$

(c) $\frac{3}{60} \times 4$

(d) $5 \times 2\frac{1}{3}$

(e) $\frac{3}{28} \times \frac{7}{30} \times \frac{15}{27}$

(f) $\frac{1}{2} \times \frac{2}{3} \times \frac{3}{4} \times \frac{4}{5} \times \frac{5}{6} \times \frac{6}{7} \times \frac{7}{8} \times \frac{8}{9} \times \frac{9}{10}$

4. Find the quotient. Leave your answers in lowest terms.

(a) $\frac{3}{7} \div 3$

(b) $\frac{1}{5} \div 2$

(c) $\frac{1}{6} \div \frac{1}{2}$

(d) $7 \div \frac{1}{3}$

(e) $\frac{3}{2} \div \frac{3}{4}$

(f) $\frac{1}{4} \div \frac{2}{3}$

(g) $\frac{12}{25} \div \frac{18}{65}$

(h) $3\frac{3}{4} \div \frac{5}{8}$

5. Simplify. Leave your answers in lowest terms.

(a) $\frac{3}{4} \div \frac{9}{2} \times \frac{2}{5}$

(b) $\left(\frac{1}{5} + \frac{1}{10}\right) \times \frac{2}{3}$

(c) $\frac{5}{9} \div \left(\frac{3}{2} - \frac{2}{3}\right)$

(d) $\left(1 + \frac{1}{2}\right)\left(1 + \frac{1}{3}\right)\left(1 + \frac{1}{4}\right)$

(e) $\dfrac{\frac{5}{3} - 1}{2 - \frac{5}{3}}$

MULTIPLYING & DIVIDING

# 10

# CHAPTER 10
## DECIMALS

### PIZZA MAN'S PLAN

I left the spy kids with a tough mixed-number multiplication question and excused myself from the café table. Once I was in the ladies' room, I shut myself into a stall and called my parents with some of my cellular plan's accrued anytime minutes.

"Sal's Pizzeria. What-a you want-a on your pizza?" Dad answered.

"It's me, Dad!"

"Oh, good, what's the news?"

"The Memorase worked. No one suspects a thing. I'm calling you from the bathroom of the Finer Diner. They really need to do a better job cleaning behind their toilets."

"You mean Orange, Taff, and the Black Mantis are alone together?"

"Yeah . . . why?"

"Get in there fast before they have a chance to compare notes! You need to get them over to the drippy cave. We have assembled a crack unit of the Coast Guard reserves. These fresh-faced recruits are tough, but they are no match for Taff and the Black Mantis. The Coast Guard reserves will just create a diversion. While our band of evil spies is preparing to overpower the reserves with their lightning-quick blend of karate, judo, and dog-paddling, your mom, my beautiful blond secretary, and I will sneak up behind each of the spies and sever their Achilles tendons with surgical scalpels! As they stagger around like stumblebums in the dewy dawn, my beautiful blond secretary will lash them together with four-inch duct tape. Only then will the world will be safe for the proliferation of U.S.-modeled, corporate-friendly representative democracies."

"Got it, Dad," I replied. "You can count on me."

I slammed out of the stall and rushed back to the table.

"Where's the fire, mate?" Kyle asked.

"We thought you fell in," Rebecca spat.

"Decimals!" I panted. "There's . . . gonna be decimals . . . on the math test . . . too. And . . . after we study for that . . . I know . . . where to get . . . more taffy . . . than you have ever imagined . . . Jake!"

# DECIMALS: THE BASICS

Like fractions, decimals are a way of expressing parts of a whole. Instead of a fraction bar, a decimal number uses a **decimal point** to separate the whole-number part from the fractional part. (When I say "fractional part," I just mean the part less than one. I'll call it the "decimal part" in this chapter to avoid confusion.) Everything to the left of the decimal point is the whole-number part; everything to the right is the decimal part.

You're probably already reasonably good with decimals: that's what we use to count money. Dollars come before the decimal point; cents come after. But I'm going to review everything anyway. Just skip whatever you already know.

## ANALYSIS OF A DECIMAL

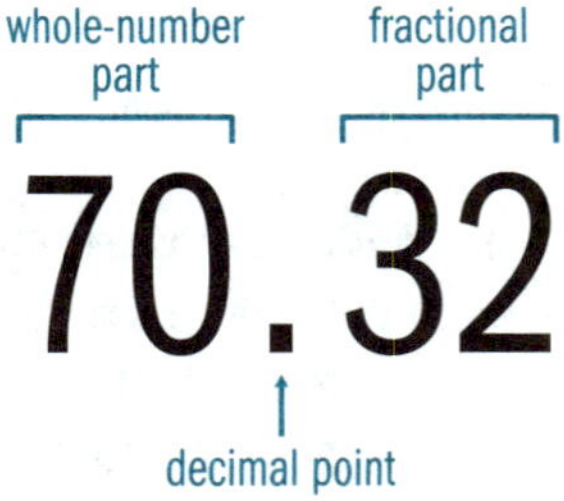


In the number $70.32$, the whole-number part is $70$. This means that $70.32$ is between $70$ and $71$.

The $.32$ is the decimal part. The decimal part is secretly a fraction whose denominator is determined by the **place values** of its digits. Because $.32$ is two digits after the decimal point, it represents thirty-two hundredths, or $\frac{32}{100}$. (More on this in the next section.)

So $70.32$ is the same thing as $70\frac{32}{100}$.

## DECIMAL PLACE VALUE

In Chapter 1 we talked about place values for digits that come before the decimal point in whole numbers. Each location *after* the decimal point also has a place value that determines how much the digit in that location is worth. Here is the number $9.723056$ with the place values of its digits labeled.

| | | | | | | | | |
|---|---|---|---|---|---|---|---|---|
| place | ones | | tenths | hundredths | thousandths | ten-thousandths | hundred-thousandths | millionths |
| | 9 | . | 7 | 2 | 3 | 0 | 5 | 6 |
| place value | 1 | | $\frac{1}{10}$ | $\frac{1}{100}$ | $\frac{1}{1000}$ | $\frac{1}{10{,}000}$ | $\frac{1}{100{,}000}$ | $\frac{1}{1{,}000{,}000}$ |

For example, the place value of the first digit after the decimal point is $\frac{1}{10}$. In $9.723056$, the digit $7$ is in the tenths' place; it's worth $\frac{7}{10}$. The place value of the third digit after the decimal point is $\frac{1}{1000}$. In $9.723056$, the digit $3$ is in the thousandths' place; it's worth $\frac{3}{1000}$.

## READING DECIMALS

Officially, we're supposed to read decimals like we read mixed numbers. So $5.6$ is "five and six tenths" and $72.125$ is "seventy-two and one hundred and twenty-five thousandths."

But this is quite a mouthful, so people often just say "five point six" and "seventy-two point one two five."

## WRITING DECIMALS

### Whole numbers

You can think of whole numbers as having an unwritten decimal point to their right. So $6$ is the same thing as $6.0$ (much the same way that $4$ is the same thing as $4\frac{0}{10}$).

### Commas?

Although we separate off groups of three digits in long whole numbers, we *don't* do that after the decimal point. Take a look:

$$3.14 \quad 2.1718281828 \quad 299{,}792.458 \quad 12{,}003{,}009.4667778$$

It's actually very rare to have a number with many digits both before and after the decimal.

### Leading zeros

When writing decimal numbers between $0$ and $1$, you can drop the zero before the decimal point. These two numbers are the same:

$$0.72 \quad \text{and} \quad .72$$

These optional zeros are called **leading zeros**. In this book, I'm going to keep writing them—they help prevent confusion between $.72$ and $72$.

### End zeros

You can pad on or drop any zeros after the last digit after the decimal point without changing the value of the number. All of these numbers have the same value:

$$2.7 \qquad 2.70 \qquad 2.700000$$

Padding with zeros is often helpful when adding or subtracting decimals.

## COMPARING DECIMALS

To figure out which of two decimal numbers is greater, first compare their whole-number parts. The one with the greater whole-number part is greater (duh!). So $5.0$ is greater than $4.9$.

To compare two decimal parts, first make sure that they have the same number of digits after the decimal point. Pad one of them with end zeros if necessary. Then compare the padded decimal parts as you would whole numbers.

For example, which is greater, $1.2$ or $1.009$?

Pad $1.2$ with zeros to make sure that it has three digits after the decimal point: $1.200$. Then compare $200$ and $009$ (which is the same thing as $9$). Since $200$ is greater, $1.2$ is greater than $1.009$.

# ADDING AND SUBTRACTING DECIMALS

Luckily for all of us, working with decimals is just like working with large numbers. The only tricky part is keeping track of the decimal point.

When adding and subtracting decimals, the key is to line up the decimal points. It can be helpful to pad with zeros to make sure that every number has the same number of digits after the decimal point. Let's just go through a bunch of examples.

## ADDING DECIMALS

### First example

What is $12.72 + 0.54$?

Line up and add. It's perfectly fine to carry across the decimal point.

$$\begin{array}{r} \overset{1}{1}2.72 \\ +\ 0.54 \\ \hline 13.26 \end{array}$$

So $12.72 + 0.54 = 13.26$.

### Second example

Sometimes, the number of digits after the decimal point changes.

What's $2.15 + 3.45$?

$$\begin{array}{r} 2.\overset{1}{1}5 \\ +\ 3.45 \\ \hline 5.60 \end{array}$$

Since $5.60$ is the same thing as $5.6$, we can say that $2.15 + 3.45 = 5.6$.

### Padding example

Be extra careful when one of the numbers has more digits after the decimal point than the other.

What's $1.302 + 0.45$?

$$\begin{array}{r} 1.302 \\ +\ 0.45\phantom{0} \\ \hline 1.752 \end{array}$$

To make the alignment easier, you can pad a zero onto the end of $0.45$ to ensure that both numbers have the same number of digits after the decimal point.

$$\begin{array}{r} 1.302 \\ +\ 0.450 \\ \hline 1.752 \end{array}$$

### Last example

Remember that whole numbers have an implicit decimal point to the right of the ones' place digit.

What's $0.004 + 21.0089 + 7$?

$$\begin{array}{r} 0.0040 \\ 21.0089 \\ +\ \ 7.0000 \\ \hline 28.0129 \end{array}$$

## SUBTRACTING DECIMALS

### First example

What's $7.23 - 2.83$?

Line up and subtract. It's fine to borrow across the decimal point.

$$\begin{array}{r} {}^{6}\ {}^{12}\phantom{0} \\ \cancel{7}.\cancel{2}8 \\ -\ 2.83 \\ \hline 4.40 \end{array}$$

So $7.23 - 2.83 = 4.40$, which is the same thing as $4.4$.

### Fewer digits in second number

If the number being subtracted has fewer digits after the decimal point, align carefully. You may or may not want to pad the second number with zeros.

What's $2.009 - 1.7$?

Here's what it looks like without padded zeros:

$$\begin{array}{r} {\scriptstyle 1\ 10\phantom{09}} \\ \not{2}.\not{0}09 \\ -\,1.7\phantom{09} \\ \hline 0.309 \end{array}$$

But it's easier to align if you pad the second number:

$$\begin{array}{r} {\scriptstyle 1\ 10\phantom{09}} \\ \not{2}.\not{0}09 \\ -\,1.700 \\ \hline 0.309 \end{array}$$

### Fewer digits in first number

If the number being subtracted has more digits after the decimal point, it's enormously helpful to pad the other number with zeros.

What's $5 - 2.41$?

Pad and work it out:

$$\begin{array}{r} {\scriptstyle 4\ \ 9\,9\,10} \\ \not{5}.\not{0}\not{0}\not{0} \\ -2.041 \\ \hline 2.959 \end{array}$$

# MULTIPLYING DECIMALS

When multiplying decimals, there's no need to worry about aligning the decimal points or padding. Joyfully ignore them while doing the multiplication. At the end, count up the total number of digits after the decimal point in *both* numbers. That's the number of digits after the decimal point that you should have in the answer.

To check your answer, estimate the product and make sure that it isn't worlds away from what you got.

To keep the next examples clean, I'll skip writing the carry digits.

### First example

$$2.25 \times 0.9 = ?$$

**Multiply:** Feel free to ignore the leading zero on $0.9$–it doesn't contribute anything to the answer.

$$\begin{array}{r} 2.25 \\ \times\, 0.9 \\ \hline 2025 \end{array}$$

**Insert decimal point:** There are two digits after the decimal point in $2.25$ and one more in $0.9$. So the answer should have *three* digits after the decimal point.

$$\begin{array}{r} 2.25 \\ \times\, 0.9 \\ \hline 2.025 \end{array}$$

So $2.25 \times 0.9 = 2.025$.

**Check:** $2.25$ is close to $2$, and $0.9$ is close to $1$. Since $2 \times 1 = 2$, the answer should be around $2$. And it is. Beautiful! If we had messed up counting the decimal points and gotten $20.25$ or $0.2025$, we'd know something was wrong–our answer would have been way off the estimate.

### Watch the end zeros!

Any end zeros that you get in the product may be insignificant in the final answer, but they *do* count as digits when you're figuring out where to place the decimal point. If you screw this up, you can get an answer that's a factor of ten off–so watch it.

What's $3.704 \times 2.5$?

**Multiply:**

$$\begin{array}{r} 3.704 \\ \times\;\; 2.5 \\ \hline 18520 \\ +\, 7408\phantom{0} \\ \hline 92600 \end{array}$$

**Insert decimal point:** Since there are three digits after the decimal point in $3.704$ and one in $2.5$, there should be *four* in the answer. The end zeros count as digits.

$$\begin{array}{r} 3.704 \\ \times\;\; 2.5 \\ \hline 18520 \\ +\, 7408\phantom{0} \\ \hline 9.2600 \end{array}$$

So $3.704 \times 2.5 = 9.26$.

**Check:** $3.704$ is close to $4$ and $2.5$ is close to $3$. So the answer should be close to $12$. And it's close enough. If we'd gotten $0.926$ or $92.6$, then we'd know there was a problem.

# MULTIPLYING AND DIVIDING BY 10

What happens when you multiply a decimal number by $10$?

Let's look at $23.45 \times 10$, using the method we just went over. Well, $2345 \times 10 = 23{,}450$. (That's from Chapter 3: multiplying by $10$ is the same thing as adding on a zero.) Since $23.45$ has two digits after the decimal point and $10$ has no digits after the decimal point, the answer must have two digits after the decimal point: $234.50$. Since that last zero is just a placeholder, we get

$$23.45 \times 10 = 234.5.$$

Did you see that? It's like the decimal point just moved over one digit.

In fact, this is exactly what always happens. Multiplying by $10$ is the same thing as moving the decimal point one digit to the right.

Take a look:

$$\begin{array}{rcccl} 72.4 & \times & 10 & = & 724 \\ 1.0009 & \times & 10 & = & 10.009 \\ 0.008 & \times & 10 & = & 0.08 \\ \$0.40 & \times & 10 & = & \$4.00 \end{array}$$

If you look at it a certain way, the adding-on-a-0 rule is just another way of formulating the moving-the-decimal-point rule. Take a look: the adding-on-a-0 rule tells us that $67 \times 10 = 670$. But since $67$ can also be thought of as $67.0$, we get exactly the same thing if we move the decimal point to the right: $67.0 \times 10 = 670$.

## MULTIPLYING BY 10, BY 100, BY 1000

What happens when you multiply a decimal number by 100? Well, multiplying by 100 is the same thing as multiplying by 10 twice. That means that multiplying by 100 is the same thing as moving the decimal point *two* digits to the right:

$$23.45 \times 100 = 2345$$
$$72.4 \times 100 = 7240$$ (because 72.4 is also 72.40)
$$1.0009 \times 100 = 100.09$$
$$0.008 \times 100 = 0.8$$
$$\$0.40 \times 100 = \$40.00$$

Similarly, multiplying by 1000 is the same thing as moving the decimal point *three* digits to the right.

$$23.45 \times 1000 = 23{,}450$$ (because 23.45 is also 23.450)
$$72.4 \times 1000 = 72{,}400$$ (because 72.4 is also 72.400)
$$1.0009 \times 1000 = 1000.9$$
$$0.008 \times 1000 = 8$$
$$\$0.40 \times 1000 = \$400.00$$

And multiplying by 10,000 is the same thing as moving the decimal point *four* digits to the right.

I bet you've already caught on to the pattern: move the decimal point as many digits as there are zeros that follow the 1 in the number you're multiplying by.

## DIVIDING BY 10, BY 100, BY 1000

Since multiplying by 10 moves the decimal point to the *right*, dividing by 10 moves the decimal point to the *left*.

$$\$720 \div 10 = \$\,72$$
$$23.45 \div 10 = 2.345$$
$$1.0009 \div 10 = 0.10009$$
$$0.008 \div 10 = 0.0008$$

Similarly, dividing by $100$ moves the decimal point two digits to the left, and dividing by $1000$ moves the decimal point three digits to the left:

$$\begin{aligned} \$720 &\div 100 = \$\,7.20 \\ 23.45 &\div 100 = 0.2345 \\ 1.0009 &\div 100 = 0.01000 \\ 0.008 &\div 100 = 0.00008 \end{aligned}$$

$$\begin{aligned} \$720 &\div 1000 = \$\,0.72 \\ 23.45 &\div 1000 = 0.02345 \\ 1.0009 &\div 1000 = 0.00100 \\ 0.008 &\div 1000 = 0.00000 \end{aligned}$$

### MULTIPLYING BY 0.1, BY 0.01, BY 0.001

Since $0.1$ is the same thing as $\frac{1}{10}$, multiplying by $0.1$ is the same thing as dividing by $10$. So to multiply by $0.1$, move the decimal point one digit to the left:

$$23.45 \times 0.1 = 2.345.$$

Similarly, to multiply by $0.01$, move the decimal point two digits to the left:

$$23.45 \times 0.01 = 0.2345.$$

And to multiply by $0.001$, move the decimal point three digits to the left:

$$23.45 \times 0.001 = 0.02345.$$

### DIVIDING BY 0.1, BY 0.01, BY 0.001

Since dividing by $0.1$ (or $\frac{1}{10}$) is the same thing as multiplying by $10$, move the decimal point to the right:

$$23.45 \div 0.1 = 234.5.$$

Similarly, to divide by $0.01$, multiply by $100$; and to divide by $0.001$, multiply by $1000$.

## DIVIDING WITH DECIMALS

When dividing decimal numbers, always check to make sure that the answer makes sense.

I'll go over all the rules for where to put the decimal point, but they're a wee bit involved. So if you prefer, you can ignore them. Instead, estimate the answer and place the decimal point accordingly. If you're careful, you can't go wrong. I'll show you how to do this with each example.

## THE EXACT QUOTIENT

We went over division with remainder in Chapter 4. Sometimes we need to know the exact quotient, however.

### First example

What's $75 \div 4$?

Start dividing through normally.

$$\begin{array}{r@{\,}r} & 18 \\ 4 \big) & \overline{75} \\ & -4\phantom{0} \\ \hline & 35 \\ & -32 \\ \hline & 3 \end{array}$$

Usually, we'd stop when there's nothing more to bring down. Whatever is left over becomes the remainder. But if we want the exact quotient, we have to press on. Place the (usually unwritten) decimal point to the right of both the number being divided and the quotient. Pad the number being divided with a zero (it's after the decimal point, so it doesn't affect anything).

$$\begin{array}{r@{\,}l} & 18. \\ 4 \big) & \overline{75.0} \\ & -4 \\ \hline & \phantom{-}35 \\ & -32 \\ \hline & \phantom{-3}3 \end{array}$$

Now we can continue dividing. Bring down that new zero, and divide through as usual.

$$
\begin{array}{r}
18.7 \\
4\overline{)75.0} \\
-4\phantom{5.0} \\
\hline
35\phantom{.0} \\
-32\phantom{.0} \\
\hline
30 \\
-28 \\
\hline
2
\end{array}
$$

There's still a remainder left, so do it all over again. Pad on another zero, bring it down, and divide:

$$
\begin{array}{r}
18.75 \\
4\overline{)75.00} \\
-4\phantom{5.00} \\
\hline
35\phantom{.00} \\
-32\phantom{.00} \\
\hline
30\phantom{0} \\
-28\phantom{0} \\
\hline
20 \\
-20 \\
\hline
0
\end{array}
$$

Finally, we're done: $75 \div 4 = 18.75$.

**Decimal point check:** If we had ignored the decimal point rules and gotten 1875 as an answer, where would we place the decimal point? We have many choices, including these:

0.1875 1.875 18.75 187.5 1875

Since 75 is close to 80, the quotient should be close to $80 \div 4 = 20$. So 18.75 is the only reasonable choice.

**Truncating the answer**

What's $38 \div 9$?

Start dividing:

$$\begin{array}{r} 4 \\ 9\overline{)38} \\ -36 \\ \hline 2 \end{array}$$

Add decimal points, pad on a zero, and divide again:

$$\begin{array}{r} 4.2 \\ 9\overline{)38.0} \\ -36\phantom{.0} \\ \hline 2\,0 \\ -1\,8 \\ \hline 2 \end{array}$$

There's still a remainder left, so do it again:

$$\begin{array}{r} 4.22 \\ 9\overline{)38.00} \\ -36\phantom{.00} \\ \hline 2\,0\phantom{0} \\ -1\,8\phantom{0} \\ \hline 2\,0 \\ -1\,8 \\ \hline 2 \end{array}$$

And there's still a remainder left, so do it *again*:

$$\begin{array}{r} 4.222 \\ 9\overline{)38.000} \\ -36\phantom{.000} \\ \hline 2\,0\phantom{00} \\ -1\,8\phantom{00} \\ \hline 20\phantom{0} \\ -18\phantom{0} \\ \hline 20 \\ -18 \\ \hline 2 \end{array}$$

And there's *still* a remainder! At some point, people get sick of it and just say that $38 \div 9$ is approximately $4.222$.

**Decimal point check:** $38$ is close to $36$, which gives $4$ when divided by $9$. So $4.222$ is a much better choice than $0.4222$ or $42.22$.

How far should you go before you give up and give an approximate answer? It all depends on context. If you're dividing $\$38$ among $9$ people, it doesn't make much sense to go past $\$4.22$, for example. But if you're a scientist working out the average speed of a ball that traveled $38.00$ meters in $9.000$ seconds, you might need more precision.

The exact quotient, by the way, is $4.222222...$, where the $2$ repeats forever. More on that when we talk about converting decimals to fractions later on in this chapter.

## DIVIDING A DECIMAL

There's nothing much new to this, so let's dig right in.

### First example

$47.1 \div 6 = ?$

Before dividing, bring up the decimal point into the quotient. Make sure to line up the first digit of the quotient carefully: since $47$ is the first group of digits that's big enough to divide by $6$, the quotient will start over the $7$.

$$\begin{array}{r} 7.8 \\ 6\overline{)47.1} \\ -42\phantom{.1} \\ \hline 5\,1 \\ -4\,8 \\ \hline 3 \end{array}$$

At this point, we already know that $47.1 \div 6$ is approximately $7.8$. But we can get a more-exact answer by padding $47.1$ with zeros after the decimal point:

$$
\begin{array}{r}
7.85 \\
6\,\overline{)\,47.10} \\
-42\phantom{.10} \\
\hline
5\,1\phantom{0} \\
-4\,8\phantom{0} \\
\hline
30 \\
-30 \\
\hline
0
\end{array}
$$

Since there's no leftover, we're done: $47.1 \div 6 = 7.85$, exactly.

**Decimal point check:** Since $47.1$ is close to $48$, the quotient should be close to $48 \div 6 = 8$. So $7.85$ makes sense.

## Another example

What's $3.29 \div 7$?

Bring up the decimal point. Since $32$ is the first group divisible by $7$, the first digit of the quotient will line up with the $2$.

$$
\begin{array}{r}
.47 \\
7\,\overline{)\,3.29} \\
-2\,8\phantom{9} \\
\hline
49 \\
-49 \\
\hline
0
\end{array}
$$

So $3.29 \div 7 = 0.47$.

**Decimal point check:** $3.29$ is close to $3$ and $7$ is close to $10$. So the quotient should be close to $3 \div 10 = 0.3$. And it is.

## A note about remainders

When you're dividing with decimals, it doesn't make sense to talk about remainders. Remainders belong to the world of whole numbers. Once you've broken through that barrier—once you start thinking about fractions and decimals—remainders are out of place. So rather than saying that $12.5 \div 5$ is equal to $2.1$ with a remainder of $2$, you should divide the remainder completely to get $12.5 \div 5 = 2.5$.

## DIVIDING BY A DECIMAL NUMBER

If the divisor has a decimal point, we make an adjustment on both numbers to eliminate it. Move the decimal point in the divisor over to the right so that it becomes a whole number. Then move the decimal point in the dividend the same number of digits to the right. Finally, divide normally.

### First easy example

What's $17.5 \div 2.5$?

**Convert:** $2.5$ has one digit after the decimal point. Move the decimal point one digit to the right to get $25$. Now do the same thing with $17.5$ to get $175$. The problem has now become $175 \div 25$.

**Divide:** I'll spare you the details, but $175 \div 25 = 7$.

So $17.5 \div 2.5 = 7$ as well.

**Check:** Does this make sense? Since $17.5$ is roughly close to $20$ and $2.5$ is roughly close to $2$, the quotient should be about $20 \div 2 = 10$. The answer $7$ looks good.

### Second easy example

What's $13 \div 0.5$?

**Convert:** $0.5$ has one digit after the decimal point. Move it over to the right to get $5$. Since $13$ is the same thing as $13.0$, moving the decimal point one digit to the right gives us $130$. So the problem is now to compute $130 \div 5$.

**Divide:** Again, you already know how to do this, so I'll just tell you: $130 \div 5 = 26$.

So $13 \div 0.5 = 26$.

**Check:** Does this make sense? Since $13$ is roughly close to $10$ and $0.5$ is roughly close to $1$, the answer should be roughly close to $10 \div 1 = 10$. The answer $26$ is close enough.

### Hard example

$$0.75\overline{)12.6}$$

**Convert:** $0.75$ has two digits after the decimal point and converts to $75$. Do the same to $12.6$, which converts to $1260$. So the problem becomes

$$75\overline{)1260}$$

**Divide,** bringing down zeros after the decimal point if necessary.

$$\begin{array}{r} 16.8 \\ 75\overline{)1260.0} \\ -75\phantom{0.0} \\ \hline 510\phantom{.0} \\ -450\phantom{.0} \\ \hline 60\ 0 \\ -\ 60\ 0 \\ \hline 0 \end{array}$$

So $12.6 \div 0.75 = 16.8$.

**Check:** Since $12.6$ is about $10$ and $0.75$ is about $1$, the answer should be close to $10 \div 1 = 10$. That's close enough to $16.8$.

# CONVERTING BETWEEN DECIMALS AND FRACTIONS

## CONVERTING DECIMALS TO FRACTIONS

This part is pretty easy. A decimal is already secretly a fraction whose denominator is determined by the place value of the last digit. The denominator is always $10$ or $100$ or $1000$ or something similar. The number of zeros should be the same as the number of digits after the decimal point of the number you're converting.

### Easy first example

Let's start with an easy one. Convert $0.8$ to a fraction.

If you know how to read decimals aloud, you're in great shape: $0.8$ is literally eight tenths, or $\frac{8}{10}$.

To finish up, reduce $\frac{8}{10}$ to lowest terms: $\frac{8}{10} = \frac{8 \div 2}{10 \div 2} = \frac{4}{5}$.
So $0.8 = \frac{4}{5}$.

### Easy second example

Express $0.375$ as a fraction in lowest terms.

The last digit of $0.375$ is in the thousandths' place, so

$$0.375 = \frac{375}{1000},$$

which reduces to $\frac{375}{1000} = \frac{375 \div 125}{1000 \div 125} = \frac{3}{8}$.

So $0.375 = \frac{3}{8}$.

### Example with a whole-number part

The easiest way to convert decimals with a whole-number part is to deal with the decimal part separately and slap the whole-number part at the end to make a mixed number—which you can then convert to an improper fraction if you like.

Let's convert $2.16$ to a mixed number.
Look at $0.16$. The last digit is in the hundredths' place, so

$$0.16 = \frac{16}{100},$$

which reduces to $\frac{4}{25}$.

That means that $2.16 = 2\frac{4}{25}$.

If we need $2.16$ as an improper fraction, we can either convert $2\frac{4}{25}$, or we can deal with $2.16$ directly:

$$2.16 = \frac{216}{100},$$

which reduces to $\frac{54}{25}$. And that's the answer.

## CONVERTING FRACTIONS TO DECIMALS

To convert a fraction into a decimal, simply divide the numerator by the denominator! Pad the numerator with a decimal point and zeros, and just keep going.

### First example

Express $\frac{3}{4}$ as a decimal:

$$\begin{array}{r} .75 \\ 4\overline{)3.00} \\ -2\,8\phantom{0} \\ \hline 20 \\ -20 \\ \hline 0 \end{array}$$

The quotient gives you the decimal: $\frac{3}{4} = 0.75$.

### Repeating example

Sometimes the division will go on and on. Let's try converting $\frac{2}{9}$.

$$\begin{array}{r} .222... \\ 9\,\overline{)2.000...} \\ -1\,8\phantom{00...} \\ \hline 20\phantom{0...} \\ -18\phantom{0...} \\ \hline 20\phantom{...} \\ -18\phantom{...} \\ \hline 2\phantom{...} \\ \ddots \end{array}$$

We keep doing the same computation over and over again. The fraction $\frac{2}{9}$ converts to $0.222...$, and the $2$s keep repeating forever. This is called a **repeating decimal**. It can be written with the ellipsis (the three dots) or with a bar over the repeating digit or digits. So

$$\frac{2}{9} = 0.222...$$

DECIMALS

or

$$\frac{2}{9} = 0.\overline{2}\,.$$

### Hard repeating example

Sometimes more than one digit is repeating. And sometimes the repetition doesn't start right away. This is all tricky stuff, so I'll just work through an example where both happen at the same time.

Express $\frac{7}{22}$ as a repeating decimal.

$$\begin{array}{r}
.31818\ldots \\
22\,\overline{)\,7.00000\ldots} \\
-6\,6\phantom{0000\ldots} \\
\hline
40\phantom{000\ldots} \\
-22\phantom{000\ldots} \\
\hline
180\phantom{00\ldots} \\
-176\phantom{00\ldots} \\
\hline
40\phantom{0\ldots} \\
\vdots\phantom{0\ldots}
\end{array}$$

This is written as $\frac{7}{22} = 0.3\overline{18}$. We can also round off and say that $\frac{7}{22}$ is approximately equal to $0.32$.

# YOUR TURN

Solutions start on page 289.

1. Name the digit in the ______ place of $123.456789012$.

   (a) tenths'
   (b) tens'
   (c) ten-thousandths'
   (d) hundreds'
   (e) hundredths'
   (f) hundred-millionths'

2. Which is greater,

   (a) $1$ or $0.1$?
   (b) $0.01$ or $0.1$?
   (c) $0.9$ or $1.0$?
   (d) $0.0879$ or $0.200$?

3. Find each sum or difference.

   (a) $3 + 1.589$
   (b) $2.19 + 3.82$
   (c) $4.05 - 1.09$
   (d) $10.3 + 4.125$
   (e) $7 - 0.45$
   (f) $9.2 - 2.09$

4. Simplify.

   (a) $678 \div 100$
   (b) $0.006 \times 10{,}000$
   (c) $38.09 \div 1000$
   (d) $870 \times 0.1$
   (e) $1.05 \div 0.01$

5. Find each product.

(a) $12 \times 0.5$

(b) $0.25 \times 17$

(c) $32 \times 0.625$

(d) $142.857 \times 0.7$

(e) $0.00202 \times 101.1$

6. Find each quotient.

(a) $36 \div 5$

(b) $1 \div 25$

(c) $7.2 \div 8$

(d) $9.1 \div 14$

(e) $5.6 \div 0.7$

(f) $1.21 \div 1.1$

(g) $30 \div 0.12$

7. Convert each decimal number to a fraction (proper or improper) in lowest terms.

(a) $0.9$

(b) $0.07$

(c) $2.4$

(d) $0.875$

(e) $1.333333...$

(f) $0.0625$

8. Convert each fraction to a decimal.

(a) $\frac{37}{100}$

(b) $\frac{3}{5}$

(c) $\frac{7}{20}$

(d) $\frac{5}{12}$

9. Match each fraction with its decimal equivalent.

| | |
|---|---|
| (a) $\frac{1}{2}$ | I. 0.25 |
| (b) $\frac{1}{3}$ | II. 0.166666... |
| (c) $\frac{1}{4}$ | III. 0.142857... |
| (d) $\frac{1}{5}$ | IV. 0.33333... |
| (e) $\frac{1}{6}$ | V. 0.5 |
| (f) $\frac{1}{7}$ | VI. 0.1 |
| (g) $\frac{1}{8}$ | VII. 0.125 |
| (h) $\frac{1}{9}$ | VIII. 0.11111... |
| (i) $\frac{1}{10}$ | IX. 0.2 |

# 11

# CHAPTER 11
## PERCENT



### MAKE TAFFY, NOT WAR

It worked! The master spies were completely duped by my vague promises of limitless taffy. After we finished with decimal-to-fraction conversion, I herded them back into my 2001 Saturn.

"There's an abandoned taffy factory at the edge of Maplewood Forest Preserve," I lied.

"Are you serious?" Rebecca asked.

"I've lived here my whole life, Becca. You've been here, what, a year? And these guys are even bigger newbies."

"I love taffy," Jake drooled. "Let's go!"

"I don't know," Rebecca hedged. "Something fishy is going on here."

"Come on, loser!" I urged. "Everyone else is coming!"

"Fine," Rebecca said. "But there's no reason to get nasty."

Good to know that peer pressure works on master spies too.

"Can we listen to our demo CD on the way, mate?" Kyle asked. "I want to see how it sounds in a 2001 Saturn."

"Absolutely," I said. I slid the CD in, and we were off. Johnny London really *was* great. I was torn. Sure, if I turned Kyle, Jake, and Rebecca in, I would be making the world a better place for corporate-friendly democracies. But I would also be robbing it of what could become one of the greatest bands of all time!

I had to decide: parents and democracies or Kyle and Johnny London. Hmm. Mom and Dad had said only that Kyle and Jake and Rebecca were rogue spies. What exactly was their take on the world? I turned down the CD player.

"Hey, guys," I said. "I have a word problem for you. It has to do with percent."

"I'm game, mate." Kyle smiled.

"Now, this is just a hypothetical. But say you were a rogue spy bent on world domination, what percentage of the world would you want to dominate quickly and why?"

"I'm a pacifist, mate," Kyle said. "I want $100$ percent of the world to be at peace. Conflicts should be resolved with epic Battle of the Bands competitions instead of wars."

"Me too," said Jake. "And I think Johnny London is already better than $96$ percent of the bands out there."

"Wait a minute!" Rebecca said. "That was a weird math question, Shanon. What's all this about rogue spies and world domination? What are you trying to get at?"

"Oh, n-nothing!" I stammered. "No, that was a real math question. A real *trick* math question. I tricked you into thinking about percent. Which we will continue to do as we drive to the abandoned taffy factory."

Whew!

# PERCENT

**Percent** are yet a third way—after fractions (Chapter 7) and decimals (Chapter 10)—of expressing parts of a whole. Percentages can be thought of as fractions with denominator $100$: in fact, the word *percent* literally means "out of a hundred." So $23$ percent is "$23$ out of a hundred," or $\frac{23}{100}$. And that's also the same thing as $0.23$.

The word *percent* is often abbreviated with the symbol %. So $23\%$ is read as "twenty-three percent."

Percentages are most often used when discussing real-world quantities. Often, a percentage is followed by the word *of* and the group we're talking about. That group represents $100\%$. For example, there are $4$ people in my car, and $75\%$ of them are rogue spies bent on world domination. Also, $50\%$ are girls. At least $25\%$ of them are drooling over taffy. A full $100\%$ are taking an after-school remedial-math class. And 0% of the people in my car speak Swahili. I think.

Just to make sure we're all on the same page, $100\%$ is the same thing as $\frac{100}{100} = 1$. So $100\%$ of a group is everything in that group: $100\%$ of the posters in my room are of British rock bands. And if you get $100\%$ on a test, you got every question right. Similarly, $0\%$ is $0$, or nothing at all. I got bad grades in $0\%$ of all the math classes I've ever taken.

Why do people use percent? Aren't fractions and decimals enough? Well, because of our number system, people have a better feeling for the denominator $100$ than for other bizarre denominators. You have to agree that it's easier to think about $83\%$ than about $\frac{296}{357}$. And nobody *really* likes the decimal point. (Do you? I sure don't. It makes me just a little anxious. What if I misplace it or something?)

# CONVERTING PERCENT TO FRACTIONS AND DECIMALS

Most of these should be really easy. You already did the hard stuff when you learned to convert between decimals and fractions.

## PERCENT TO DECIMALS

Drop the % sign and move the decimal two places to the *left*. (That's the same thing as dividing by $100$. Take another look at Chapter 10 to see why.)

$47\%$ is the same as $0.47$.

If you're converting a quantity less than $10\%$, you'll have to slap on a leading zero:

$8\%$ is the same as $0.08$.

This works even if the percentage has a decimal point:

$29.5\%$ is the same as $0.295$;

$2.3\%$ is the same as $0.023$.

Here's a tricky one:

$0.03\%$ is the same as $0.0003$.

## DECIMALS TO PERCENT

Go the other way! Move the decimal point two places to the *right* (that's multiplying by $100$) and slap on a % sign. Take a look at these examples:

$0.71$ is the same as $71\%$

$0.04$ is the same as $4\%$

$0.349$ is the same as $34.9\%$

$0.0082$ is the same as $0.82\%$

## PERCENT TO FRACTIONS

Take off that silly percent sign and write the percentage as a fraction over the denominator $100$. Then reduce if you must (go to Chapter 7 for a review of reducing fractions).

So $27\%$ is the same thing as $\frac{27}{100}$, which doesn't need to be reduced.

But $40\%$ converts to $\frac{40}{100}$, which reduces to $\frac{2}{5}$.

## FRACTIONS TO PERCENT

You can do this in one of two ways. You can convert the fraction to a decimal and then move the decimal point over two places to the right. So to express $\frac{3}{8}$ as a percentage, do the division:

$$\begin{array}{r} 0.375 \\ 8\overline{)3.000} \\ -2\,4\phantom{00} \\ \hline 60\phantom{0} \\ -56\phantom{0} \\ \hline 40 \\ -40 \\ \hline 0 \end{array}$$

So $\frac{3}{8} = 0.375$.

And $0.375$ is the same thing as $37.5\%$.

Alternatively, you can multiply the fraction by $100$, then reduce it down to a mixed number or a decimal. Let's see how this works for $\frac{3}{8}$:

Multiply by $100$:

$$\frac{3}{8} \times 100 = \frac{3 \times \overset{25}{\cancel{100}}}{\underset{2}{\cancel{8}}} = \frac{75}{2}$$

Reduce to a mixed number or a decimal: $\frac{75}{2}$ is the same thing as $37\frac{1}{2}$ or $37.5$.

So $\frac{3}{8}$ is $37\frac{1}{2}\%$ or $37.5\%$.

These two methods are essentially the same: the first method simplifies and then multiplies by 100, and the second method multiplies by 100 and then simplifies.

## COMMON PERCENTAGES

There are a few percentages that show up again and again, so I'm going to list them here along with their fractional and decimal conversions. It's not like you can't survive without memorizing them, but they're sure nice to know.

| Fraction | Decimal | Percent |
|---|---|---|
| 0 | 0 | 0% |
| $\frac{1}{8}$ | 0.125 | 12.5% |
| $\frac{1}{6}$ | 0.1666... | 16.7% |
| $\frac{1}{5}$ | 0.2 | 20% |
| $\frac{1}{4}$ | 0.25 | 25% |
| $\frac{1}{3}$ | 0.3333... | 33.3% |
| $\frac{2}{5}$ | 0.4 | 40% |
| $\frac{1}{2}$ | 0.5 | 50% |
| $\frac{2}{3}$ | 0.6666... | 66.7% |
| $\frac{3}{4}$ | 0.75 | 75% |
| $\frac{4}{5}$ | 0.8 | 80% |
| $\frac{5}{6}$ | 0.8333... | 83.3% |

| Fraction | Decimal | Percent |
| --- | --- | --- |
| $\frac{7}{8}$ | 0.875 | 87.5% |
| 1 | 1 | 100% |

# WORKING WITH PERCENT

There are three common types of percent problems, and they all have the same setup. There is a "whole" (representing 100%), a percentage, and a "part" (representing the percentage). We know two of these three numbers—whole, part, and percentage—and we're asked to find the third.

Let's go over how to do all three types, from easiest to hardest.

Also, a lot of the percentage questions that math students have to answer come in the form of word problems. So we'll do a few of those.

One last thing: in all but the trickiest of problems, the "whole" number will be larger than the "part" number.

## WHAT IS ___% OF ___?

We know the whole and the percentage, and have to find the part.

Here's an example: I've noticed that 75% of the students in our after-school math class can't add fractions. If there are 32 kids in the class, how many of them couldn't compute $\frac{1}{2} + \frac{2}{3}$ if their life depended on it?

Or more simply, what's 75% of 32 students? The 32 is the "whole" and 75% is the percentage. What's the "part"?

We have to convert this question to something mathy. We already know that 75% is the same thing as 0.75. And in situations like this, the word *of* may as well be replaced with the word *times* (as in multiplication). So we can rewrite the question:

What is 75% of 32 students?

↓ ↓ ↓

What is 0.75 × 32 ?

Now the computation is easy: $0.75 \times 32 = 24$. So 75% of 32 students is 24 students—that's the number of my classmates who can't add fractions. (And Kyle is one of the worst.)

Let's try just one more example. My monthly allowance is $56. Last month, I spent exactly 42% of my allowance at the Finer Diner. How much money do I drop on food every month?

Again, first get at the core of the question: What is 42% of 56?

Then convert: 42% of 56 is the same thing as

$$0.42 \times 56,$$

and that's easy to compute:

$$0.42 \times 56 = 23.52.$$

So I spent $23.52 Finer-Dining. No wonder my thighs are so plump.

## WHAT PERCENT OF ___ IS ___?

Now we know the whole and the part, we have to find the percentage.

Kyle has 25 different guitar picks, and 4 of them are blue. What percentage of Kyle's guitar picks are blue?

In other words, what percent of 25 is 4?

Because 25 is the *total number* of picks, it's the whole: it represents 100%. The smaller number 4 is the part. We need to know what portion of 25 the 4 blue picks are—in other words, how much 4 out of 25 is.

So we divide:

$$\frac{4}{25} = 0.16$$

and then convert the decimal to a percentage: 0.16 is the same thing as 16%.

So 16% percent of Kyle's picks are blue.

You can easily check your work: 16% of 25 is the same thing as $0.16 \times 25$, which is 4. Perfect.

The same question can be phrased slightly differently, with the order of the two numbers inverted. Instead of "What percent of 25 is 4?", it can be formulated as "4 is what percent of 25?" Don't be fazed! The number that represents 100% will usually have an *of* just before it.

Here's an example: 26 is what percent of 65?

The number 65 represents 100%, so divide:

$$\frac{26}{65} = 0.4.$$

So 26 is 40% of 65.

## ___ IS ___% OF WHAT NUMBER?

We know the part and the percentage, and have to find the whole. This is the trickiest kind of problem.

To solve, divide the part by the percentage converted to a decimal.

Example: 20 is 8% of what number?

Convert 8% to a decimal to get 0.08.

Then divide: $\frac{20}{0.08} = 250.$

So 20 is 8% of 250.

**Check:** $250 \times 0.08 = 20$. Good.

Essentially, it works because when you multiply the whole by the percentage, you get the part. In other words, the whole is the number that, when multiplied by the percentage, gives you the part. So to get the whole back, you have to *divide* the part by the percentage.

## TOTAL = 100%

There is one more type of percent problem that comes up often. It's completely different from the previous three types, but it's extremely easy—so I thought I'd slip it in.

The only thing to keep in mind as you're doing these problems is that the whole (group of people, or set of picks, or number of shoes in the world) is *always* 100%.

Say 62% of the kids in my school are girls. How many are boys?

Well, the whole school population is 100%. If 62% are girls, then 100% – 62% = 38% are boys. Easy!

Let's do another. Jake collects taffy. His taffy comes in three flavors—raspberry, boysenberry, and mulberry. Only $9\%$ of the taffy he has is boysenberry flavored, but a whopping $62\%$ is mulberry. What percentage of his taffy tastes like raspberries?

Again, his whole taffy collection is $100\%$. Boysenberry and mulberry together make $62\% + 9\% = 71\%$. So there's $100\% - 71\% = 29\%$ left for raspberry.

Sometimes, there might be extra information in the problem. Don't be led astray!

For example, Becca flips a coin $352$ times. The coin lands heads up $54\%$ of the time. What percentage of her flips are tails?

You might be tempted to figure out that she flipped $0.54 \times 352 = 190$ heads . . . but who cares? All you need to know is the *percentage* of tails flips. Since all the flips make $100\%$, you know that $100\% - 54\% = 46\%$ of them are tails. It looks like this wasn't a very fair coin.

# YOUR TURN

Solutions start on page 296.

1. Convert each percentage to a decimal.

   (a) $30\%$

   (b) $0\%$

   (c) $4\%$

   (d) $110\%$

   (e) $0.5\%$

   (f) $1\frac{1}{4}\%$

2. Convert each fraction or decimal to a percentage.

   (a) $0.37$

   (b) $1.02$

   (c) $0.006$

   (d) $4.5$

   (e) $\frac{7}{10}$

   (f) $\frac{3}{16}$

3. (a) Find $50\%$ of $12$.

   (b) Find $120\%$ of $15$.

   (c) What is $6\%$ of $72$?

   (d) What number is $1.5\%$ of $200$?

4. (a) What percent of $65$ is $13$?

   (b) $80$ is what percent of $40$?

   (c) $40$ is what percent of $80$?

   (d) What percent of $2$ is $1.25$?

5. (a) $300$ is $30\%$ of what number?

   (b) $60$ is $16\%$ percent of what number?

   (c) $27$ is $90\%$ of what number?

   (d) $36$ is $240\%$ of what number?

6. On Tuesday, Mr. Schnitzel–Maplewood High's German teacher–gave a test to all $68$ of the students in his four classes. That night he managed to grade $17$ of the tests before falling asleep in front of the television. What percentage of the tests remains ungraded in the wee hours of Wednesday morning?

7. Dan Visel is the smartest kid at Maplewood High. He spends $37\%$ of his intellectual brain energy on studying biology, physics, and chemistry; $20\%$ of his intellectual brain energy on reading fiction books; $16\%$ on learning French, German, and Russian; $9\%$ on perusing scholarly articles on psychology; and the rest on art appreciation. What percentage of Dan Visel's intellectual brain energy goes to appreciating art?

8. My friend Laurie has $20$ pets, of which $70\%$ are parrots and the rest are dogs. Also, $33\frac{1}{3}\%$ of the dogs are poodles and the rest are Labrador retrievers. How many Labs does Laurie have?

# 12

# CHAPTER 12
## EXPONENTS AND POWERS



### BATTLE OF THE BANDS

A mile away from my parents' humid spy cave at the edge of Maplewood Forest Preserve, I realized I couldn't do it! I couldn't watch as my only friends in the world had their Achilles tendons severed and wrapped in four-inch duct tape. I pulled over.

"What are you doing?" Jake yelled. "Taffy ahead!"

"There is no taffy, Jake," I said.

"I knew it!" Rebecca said. "'Fess up, Shanon!"

"Okay, I know that you are Agent Orange, Agent Taff, and the Black Mantis!" I admitted. "So go ahead and Mongolian torture me or wrap me in duct tape or do whatever you spies do to helpless victims such as myself."

"How did you find out?" Kyle asked.

"It's not important," I said. "What is important is that you guys change your ways. There is a small army of Coast Guard reserves waiting for you at the edge of those trees!"

"I agree with Shanon," said Kyle. "I'm sick of spying. I want to rock!"

"Me too!" Jake agreed. "And eat taffy."

"Weaklings!" Rebecca seethed. "We are *this* close to achieving our goal of world domination and the end of armed conflict in favor of a Battle of the Bands series that will confer upon the winner of the dual titles World's Greatest Rock 'n' Roll Band *and* Ruler of the Universe!"

"Is that really what you're working for?" I asked.

"Yes!" said Kyle. "That's why Johnny London is so good. We have to be to win consecutive Battle of the Bands competitions and rule the Earth and its neighboring planets!"

I thought for a minute.

"But here's the thing," I argued. "It seems to me that your Battle of the Bands contest is going to escalate into mutually assured nuclear annihilation."

"Huh?" said Kyle and Jake.

"Huh?" said the Black Mantis evilly.

"Let me explain," I said. "But first, we have to go over exponents stuff. Exponential growth is the key step in my fascinating theory."

* * * * *



Repeated addition has another name–it's called multiplication. Repeated multiplication gets its own name too: it's called **raising to powers** or **exponentiation**. Let's start with the easy stuff.

# SQUARES

The **square** of a number is its product with itself.

**Squaring**–multiplying a number by itself–is denoted with a little raised $2$ to the right of the number. Like this:

$$3^2.$$

So $3^2$ is equal to $3 \times 3$, or $9$.

The expression $3^2$ is usually read as "three squared." Occasionally people also say "three raised to the second power." But more on that later.

That little $2$ is called the **exponent**. It indicates that *two* $3$s are being multiplied together.

## PERFECT SQUARES

Squares of whole numbers are called **perfect squares**. It's nice to be able to recognize them quickly, so here's a list of the first few perfect squares.

$$
\begin{array}{ccccc}
0^2 & = & 0 \times 0 & = & 0 \\
1^2 & = & 1 \times 1 & = & 1 \\
2^2 & = & 2 \times 2 & = & 4 \\
3^2 & = & 3 \times 3 & = & 9 \\
4^2 & = & 4 \times 4 & = & 16 \\
5^2 & = & 5 \times 5 & = & 25 \\
6^2 & = & 6 \times 6 & = & 36 \\
7^2 & = & 7 \times 7 & = & 49 \\
8^2 & = & 8 \times 8 & = & 64 \\
9^2 & = & 9 \times 9 & = & 81 \\
10^2 & = & 10 \times 10 & = & 100
\end{array}
$$

## SQUARE?

Why does the word *square* mean the product of a number with itself? It makes sense if you think a little bit about geometry. The area of a square with side length of $5$ is $5 \times 5 = 25$. So its area is $5^2$, or five *squared*. (Get it?)

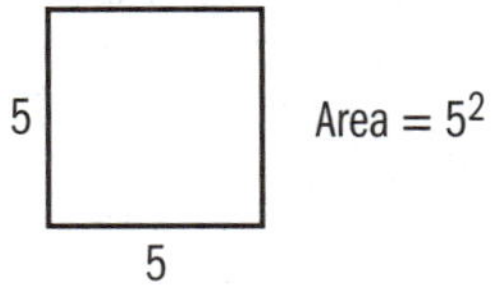

# CUBES

The **cube** of a number is its product with itself twice—three copies of the number multiplied together.

**Cubing** a number is denoted with a little $3$ to the right of the number:

$$4^3 = 4 \times 4 \times 4.$$

The **exponent** $3$ means that *three* $4$s are being multiplied together. The expression $4^3$ is read as "four cubed" or "four raised to the third power."



## PERFECT CUBES

Cubes of whole numbers are called **perfect cubes**. Perfect cubes don't come up as often as perfect squares, but I'll list a few of them:

$$\begin{aligned}
0^3 &= 0 \times 0 \times 0 = 0 \\
1^3 &= 1 \times 1 \times 1 = 1 \\
2^3 &= 2 \times 2 \times 2 = 8 \\
3^3 &= 3 \times 3 \times 3 = 27 \\
4^3 &= 4 \times 4 \times 4 = 64 \\
5^3 &= 5 \times 5 \times 5 = 125
\end{aligned}$$

## CUBE?

Again, the origins of the term *cube* can be explained with geometry. A cube with side length $2$, for example, will have volume $2 \times 2 \times 2$, or $2^3$.

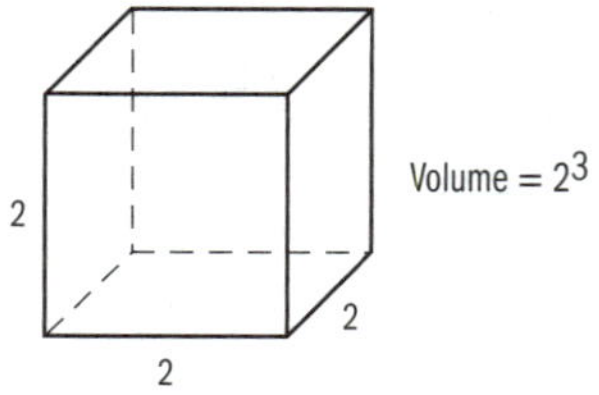

# EXPONENTS AND POWERS

Squaring and cubing are both special cases of repeated multiplication, or raising to powers. Squaring a number is raising it to the *second* power. Cubing a number is raising it to the *third* power. You can also raise numbers to the fourth, and the fifth, and higher powers.

A lot of this stuff is a repeat of what we just talked about for squares and cubes, so read fast.

## WRITING AND TALKING

The expression

$$2^6$$

means six $2$s multiplied together, or $2 \times 2 \times 2 \times 2 \times 2 \times 2$.

So $2^6 = 64$.

That $2$ is called the **base**—that's the number that gets multiplied. The little raised $6$ is the **exponent**; it tells you how many times the base is multiplied by itself.

The expression $2^6$ is read as "two raised to the sixth power," or simply "two to the sixth." We can also say that $64$ is a **power** of $2$. (This is analogous to saying that $24$ is a multiple of $8$.)

## POWERS

Sometimes people are interested at looking at all the powers of a particular number. For example, the powers of $2$ play an important role in computers and other digital technology. They come up pretty often, so here are the first few:

$$2^1 = 2 = 2$$
$$2^2 = 2 \times 2 = 4$$
$$2^3 = 2 \times 2 \times 2 = 8$$
$$2^4 = 2 \times 2 \times 2 \times 2 = 16$$
$$2^5 = 2 \times 2 \times 2 \times 2 \times 2 = 32$$
$$2^6 = 2 \times 2 \times 2 \times 2 \times 2 \times 2 = 64$$
$$2^7 = 2 \times 2 \times 2 \times 2 \times 2 \times 2 \times 2 = 128$$

$$
\begin{aligned}
2^8 &= 2\times2\times2\times2\times2\times2\times2\times2 &&= 256\\
2^9 &= 2\times2\times2\times2\times2\times2\times2\times2\times2 &&= 512\\
2^{10} &= 2\times2\times2\times2\times2\times2\times2\times2\times2\times2 &&= 1024
\end{aligned}
$$

The fact that $2^{10}$ is pretty close to $1000$ actually turns out to be pretty convenient in naming computer memory. The prefix *kilo-* normally means "one thousand"—so a kilometer is a thousand meters. But computer memory comes in packets of powers of $2$. (That's why it's called a *binary* system: *bi-* means "two," as in *bicycle* and *bisexual*.) So computer scientists use *kilo-* to mean "$1024$": a kilobyte is actually $1024$ bytes, which is close enough to $1000$.

The powers of most other numbers don't come up as often. What's important to keep in mind is that powers grow really fast—so $8^7$ is much, much bigger than, say, $8 \times 7$. Here are the first few powers of $3$.

$$
\begin{aligned}
3^1 &= 3 &&= 3\\
3^2 &= 3\times3 &&= 9\\
3^3 &= 3\times3\times3 &&= 27\\
3^4 &= 3\times3\times3\times3 &&= 81\\
3^5 &= 3\times3\times3\times3\times3 &&= 243
\end{aligned}
$$

The powers of $3$ grow much faster than the powers of $2$. For example, $3^5 = 243$ is just a little bit smaller than $2^8 = 256$.

Here are a few powers of $4$:

$$
\begin{aligned}
4^1 &= 4 &&= 4\\
4^2 &= 4\times4 &&= 16\\
4^3 &= 4\times4\times4 &&= 64\\
4^4 &= 4\times4\times4\times4 &&= 256\\
4^5 &= 4\times4\times4\times4\times4 &&= 1024
\end{aligned}
$$

Did you notice that powers of $4$ are all also powers of $2$? That's because $4$ itself is a power of $2$.

## POWERS OF 10

The powers of $10$ are easy to recognize: their first digit is $1$ and all the rest are 0s. The number of zeros that follow the $1$ corresponds to the exponent. So $1000$ is $10^3$ and $10,000$ is $10^4$.

The powers of $10$ form the basis of the number system that we use: each place value is a power of $10$. That's why our number system is called the *decimal* system: *dec-* means "ten."

Here are the first few powers of $10$. Because they're so important to the way we count, quite a few of them have special names.

| | | | |
|---|---|---|---|
| $10^1$ | = | 10 | ten |
| $10^2$ | = | 100 | one hundred |
| $10^3$ | = | 1000 | one thousand |
| $10^4$ | = | 10,000 | ten thousand |
| $10^5$ | = | 100,000 | one hundred thousand |
| $10^6$ | = | 1,000,000 | one million |
| $10^7$ | = | 10,000,000 | ten million |
| $10^8$ | = | 100,000,000 | one hundred million |
| $10^9$ | = | 1,000,000,000 | one billion |

## POWERS OF 0 AND 1

Since any number times 0 is 0, all the powers of 0 are the same. They're all 0:

$0^1 = 0$ and $0^{11} = 0$ and $0^{456} = 0$.

And since $1$ times $1$ is always $1$, all the powers of $1$ are equal to $1$:

$1^1 = 1$ and $1^7 = 1$ and $1^{89} = 1$.

Isn't that nice?

EXPONENTS & POWERS

## THE FIRST POWER

Any number to the first power is itself.

So

$$0^1 = 0 \quad \text{and} \quad 12^1 = 12 \quad \text{and} \quad 4092^1 = 4092.$$

This makes perfect sense: the exponent $1$ tells us that the base appears only once in the final product.

## ZERO POWER

This might seem a little weird, but it ends up being extremely convenient.

Any number to the zero power is $1$.

So

$$1^0 = 1 \quad \text{and} \quad 29^0 = 1 \quad \text{and} \quad 10{,}0000^0 = 1.$$

### Why this isn't ridiculous

Well, for one thing, $29^0$ means zero $29$s multiplied together, and the "nothing" product is $1$. It's similar to the fact that $29 \times 1$ is just $29$ again: the number $1$ doesn't affect anything in multiplication.

Moreover, having the zero power of anything be $1$ ends up being enormously useful. (Feel free to skip the rest of this section if you don't care.)

Compare consecutive powers of, say, $3$. For example, $3^3 = 27$ is three times as big as the previous power, $3^2 = 9$. And $3^4 = 81$ is three times bigger than $3^3 = 27$. Even $3^2 = 9$ is three times the next smaller power, $3^1 = 3$. It's a nice pattern.

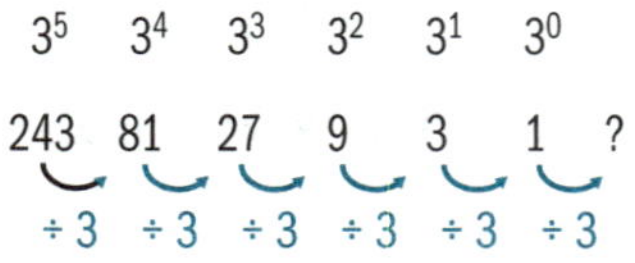


Wouldn't it be nice to continue it? That's what math people thought, so they *defined* $3^0$ to be equal to $1$ to force it to fit the pattern: now $3^1$ is three times bigger than $3^0$. In some sense, it was an arbitrary decision—you can't really multiply a number by itself zero times. But in another sense, it was a really good decision: now all the powers of $3$ have the nice property that they're three times as big as the next smaller power.

### Important exception

Oh, yeah—there's one exception to the zero-power rule.

$$0^0 \text{ is } \textbf{\textit{not defined.}}$$

It's just like the no-dividing-by-zero rule. You just can't raise zero to the zeroth power. It's impossible.

You can start to see that there's a problem if you think about it. On the one hand, $0$ to any power is supposed to be $0$. On the other hand, anything to the zero power is supposed to be $1$. Since $0^0$ is both, there's a real contradiction.

But the real problem is even more devastating. Think back on how we figured out what $3^0$ has to be: we wanted $3^1$ to be three times bigger than $3^0$, so we took $3^1$ and divided it by $3$. If we try to do the same thing to $0^1$ . . . well, you see the problem. Dividing by 0 is a big no-no. So don't do it.



# MULTIPLYING AND DIVIDING POWERS: SAME BASE

What do you get when you multiply $2^3$ by $2^4$?

Well, $2^3 = 8$ and $2^4 = 16$, so

$$2^3 \times 2^4 = 8 \times 16 = 128.$$

But $128$—check it out in the table above—is also a power of $2$: it's $2^7$. And $7 = 3 + 4$.

Coincidence? I think not.

Take a look: $2^3$ is $2 \times 2 \times 2$. And $2^4$ is $2 \times 2 \times 2 \times 2$. So

$$\begin{aligned} 2^3 \times 2^4 &= 2 \times 2 \times 2 \qquad \times \qquad 2 \times 2 \times 2 \times 2 \\ &= 2 \times 2 \times 2 \times 2 \times 2 \times 2 \times 2 \times 2 \times 2, \end{aligned}$$

which is exactly $2^7$.

## MULTIPLYING POWERS

To multiply two powers that have the same base, simply add their exponents.

So

$$2^3 \times 2^4 = 2^{3+4} = 2^7.$$

This property works really nicely with the zero-power rule. Take a look:

$$3^5 \times 3^0 = 3^{5+0} = 3^5,$$

and that's the exact same thing you get if you actually do it out:

$$3^5 \times 3^0 = 243 \times 1 = 243 = 3^5.$$

## DIVIDING POWERS

To find the quotient of two powers that have the same base, simply find the difference of their exponents.

For example,

$$5^4 \div 5^3 = 5^{4-3} = 5^1.$$

This rule makes tons of sense if you write everything out:

$$5^4 \div 5^3 = \frac{\cancel{5} \times \cancel{5} \times \cancel{5} \times 5}{\cancel{5} \times \cancel{5} \times \cancel{5}} = 5.$$

This property, too, works beautifully with the zero-power rule (that's why the zero-power thing is so helpful).

For example, we certainly want

$$4^3 \div 4^3$$

to be equal to $1$ (because any number divided by itself should give $1$). And that's exactly what we get–

$$4^3 \div 4^3 = 4^{3-3} = 4^0 = 1$$

–all because $4^0 = 1$.

### WATCH OUT!

These two properties work *only* when the bases of the powers are the same. For example, there's no way to rewrite

$$2^3 \times 5^4$$

without multiplying everything out.

# MULTIPLYING AND DIVIDING POWERS: SAME EXPONENT

What's $2^2 \times 3^2$?

Well, $2^2 = 4$ and $3^2 = 9$, so

$$2^2 \times 3^2 = 4 \times 9 = 36.$$

But $36$ is also the square of $6$. Hmm.

Let's trace what happens when we write everything out:

$$2^2 \qquad \times \qquad 3^2$$

expands to

$$2 \times 2 \qquad \times \qquad 3 \times 3,$$

which we can regroup as

$$2 \times 3 \qquad \times \qquad 2 \times 3,$$

which simplifies to

$$6 \qquad \times \qquad 6,$$

or

$$6^2.$$

## PRODUCT OF POWERS

The product of two powers is the power of the product.

So

$$2^2 \times 3^2 = (2 \times 3)^2.$$

This works for any power and can be quite helpful. For example,

$$2^7 \times 5^7 = (2 \times 5)^7,$$

which is easy to compute because $2 \times 5 = 10$, and $10$ is an easy number to exponentiate.

## QUOTIENT OF POWERS

The quotient of two powers is the power of their quotient.

So

$$12^5 \div 3^5 = (12 \div 3)^5.$$

This property is more commonly written in fraction form:

$$\frac{12^5}{3^5} = \left(\frac{12}{3}\right)^5.$$

## HEADS UP!

These two properties only work when the exponents are the same. For example, there's no way to rewrite

$$2^3 \times 5^4$$

as a power of $2 \times 5$.

Also, these properties only work for sums and products. They absolutely *do not* work for sums and differences. Take a look at the next section.

# ADDING AND SUBTRACTING POWERS

Sorry! There aren't really any helpful properties. If you come across something like

$$3^2 + 7^2$$

or even

$$2^4 - 2^3,$$

there's nothing much you can do except the math.

Oh—and by the way, you *cannot* simplify $3^2 + 7^2$ as $(3 + 7)^2$. *It just doesn't work.* Try it yourself.

# POWER OF A POWER

One last property: To raise a power to another power, multiply the exponents.

So

$$(2^3)^2 = 2^{3 \times 2}.$$

Check it out:

$$(2^3)^2 = 8^2 = 64,$$

which is the same thing as $2^{3 \times 2}$, or $2^6$.

# PROPERTIES OF EXPONENTS

Whew! We just rushed through a ton of properties and rules of exponents. So here's a quick summary.

EXPONENTS & POWERS

## ZERO POWER RULE

Any number to the zeroth power is $1$:

$$7^0 = 1.$$

Exception: $0^0$ is undefined.

## MULTIPLYING POWERS

If the bases of two powers are the same, then to multiply, add their exponents:

$$2^3 \times 2^8 = 2^{3+8}.$$

## DIVIDING POWERS

If the bases of two powers are the same, then to divide, subtract their exponents:

$$3^7 \div 3^4 = 3^{7-4}.$$

## RAISING POWERS TO POWERS

To raise a power to a power, multiply the exponents:

$$\left(2^4\right)^3 = 2^{4\times3}.$$

## RAISING A PRODUCT TO A POWER

The power of a product is the product of the powers:

$$(3 \times 4)^5 = 3^5 \times 4^5.$$

## RAISING A QUOTIENT TO A POWER

The power of a quotient is the quotient of the powers:

$$\left(\frac{2}{3}\right)^4 = \frac{2^4}{3^4}.$$

# POWERS OF FRACTIONS AND DECIMALS

We've been talking about whole numbers this entire time, but any number can be squared or cubed or raised to other powers. And all of the properties of exponents that we just reviewed work for decimals and fractions as well as for whole numbers.

Squares come up a little more often than other powers, so let's look a little closer.



## SQUARES OF DECIMALS

For example,

$$1.2^2 = 1.2 \times 1.2 = 1.44.$$

The square of a number less than $1$ will be smaller than the number, so don't be perturbed. Keep careful track of the decimal point:

$$0.3^2 = 0.3 \times 0.3 = 0.09.$$

## SQUARES OF FRACTIONS

When squaring a fraction, square both the numerator and the denominator—that's the power-of-a-quotient rule in action:

$$\left(\frac{2}{3}\right)^2 = \frac{2}{3} \times \frac{2}{3} = \frac{4}{9}.$$

Be sure to include the parentheses around the fraction being squared. If you leave it out, you run a very real risk of confusing $\left(\frac{2}{3}\right)^2$ and $\frac{(2)^2}{3}$.

Again, the square of a proper fraction will be smaller than the fraction itself. For example, above, $\frac{4}{9}$ is smaller than $\frac{2}{3}$.

# YOUR TURN

Solutions start on page 298.

1. (a) Find $13^2$.

   (b) Find the square of $28$.

   (c) What is the cube of $7$?

   (d) What is $5^4$?



2. Use the rules of exponents to rewrite each expression as a power of $6$.

   (a) $6^2 \times 6^7$

   (b) $6^{11} \div 6^{7-2}$

   (c) $\dfrac{6^{25}}{6^{17}}$

   (d) $(6^7)^9$

   (e) $2^5 \times 3^5$

   (f) $\dfrac{27^{14} \times 2^{14}}{3^9 \times 3^{19}}$

3. Use the rules of exponents to rewrite each expression as a power of $4$.

   (a) $2^{12}$

   (b) $16^5$

   (c) $64^8$

   (d) $8^{22}$

   (e) $4^{10} - 3 \times 4^9$

   (f) $2^7 + 2^7$

4. Find the square of each number.

(a) $1.5$

(b) $0.9$

(c) $0.18$

(d) $\frac{2}{5}$

(e) $2\frac{1}{3}$

5. (a) How many digits are there in $10^7$?

(b) How many digits are there in $100000^{2^4+3}$?

EXPONENTS & POWERS

# 13

# CHAPTER 13
## SQUARE ROOTS

### PROFITS AND LOSSES

"I get it now," Rebecca said. "Rock 'n' roll isn't the answer to the world's problems."

"So we *shouldn't* overthrow the U.S. government and institute a winner-takes-all Battle of the Bands competition?" Jake clarified.

"Not according to my escalation theory," I said.

"But . . . can we still get taffy?" Jake whined.

"Of course!" Kyle said. "Come 'ere, you lug!" Kyle gave Jake a friendly noogie.

"Now all you guys have to do is convince my parents and the Coast Guard reserves that you're no longer plotting a major coup," I reminded them. "I'll give them a call." I speed-dialed my dad.

"Sal's Pizzeria. What-a you want-a on-a your pizza?" Dad answered.

"It's me, Dad."

"Where are you?"

"Like you don't know! I'm sure you've been tracking us."

"Okay, why are you guys stopped 0.7 miles from the Maplewood Forest Preserve?"

"They want to negotiate."

"What!? You tipped them off?"

"Dad, they may be enemy spies, but they're also my friends."

"I'm putting your mother on," Dad said.

"What's going on, sweetie?" Mom asked.

"Kyle, Jake, and Rebecca want to negotiate. They have given up their plans for overthrowing the U.S. government in a massive Battle of the Bands. Instead, Johnny London simply wants a guaranteed recording contract with the Pfufferfish label."

"We could arrange that," Mom said. "Anything else?"

"Johnny London, managed by the Black Mantis—I mean, Rebecca—wants to fashion their own intricate profit-sharing contract, whereby a certain percentage of revenue from sales of Johnny London's music and music-related paraphernalia will go to peace- and rock-'n'-roll-promoting charities."

"Oh, I don't know, dear. Let me get your dad's beautiful blond secretary on the line. She's better with numbers than I am."

"Better yet, have her ride her bike out to my Saturn. We're at the scenic overlook, 0.7 miles from the Maplewood Forest Preserve. We'll hammer everything out here."

"Okay, dear."

"And Mom?"

"Yes, sunny?"

"No Coast Guard reserves! She comes alone, or the deal is off."

"Is that any way to talk to your mother!?"

"Mom . . ."

"Okay. Bye. I love you, honey."

"I love you, too," I said, and hung up. "They have agreed in principle to your demands!"

"Yeah, yeah," said Rebecca. "I'm already crunching some numbers. If only I knew more about square roots! I think I could really write a contract that would guarantee our charities maximum dollars."

* * * * *

The operations that we've been studying all come in pairs. Subtraction undoes the effects of addition. Division undoes multiplication. **Extracting a root** undoes raising to a power.

# SQUARE ROOTS

The **square root** of a number is a number that, when multiplied by itself, gives the original number. For example, $3$ is the square root of $9$, because $3 \times 3$ is $9$.

When squared, the square root of a number gives the number back again. So $3$ is the square root of $9$ because $3^2 = 9$. Similarly, $5$ is the square root of $25$ because $5^2 = 25$.

## RADICALS AND RADICANDS

Taking a square root is denoted with a funny V-shaped sign:

$$\sqrt{9} = 3.$$

The $\sqrt{\ }$ sign is officially called the **radical sign**, but people often just say "square root sign."

The expression $\sqrt{9}$ is read as "the square root of nine."

The number under the square root sign (the 9 in $\sqrt{9}$, for example) is called the **radicand**. This is a pretty fancy word, and you won't hear it too often.

## SQUARE ROOTS OF PERFECT SQUARES

Taking the square root of a perfect square is easy enough. You just have to know what it's the perfect square *of*. Take a look at these perfect squares and their square roots. They should all be very, very familiar.

$$\begin{array}{lclcl} \sqrt{0} & = & 0 & \text{because} & 0^2 & = & 0 \\ \sqrt{1} & = & 1 & \text{because} & 1^2 & = & 1 \\ \sqrt{4} & = & 2 & \text{because} & 2^2 & = & 4 \\ \sqrt{9} & = & 3 & \text{because} & 3^2 & = & 9 \\ \sqrt{16} & = & 4 & \text{because} & 4^2 & = & 16 \\ \sqrt{25} & = & 5 & \text{because} & 5^2 & = & 25 \\ \sqrt{36} & = & 6 & \text{because} & 6^2 & = & 36 \\ \sqrt{49} & = & 7 & \text{because} & 7^2 & = & 49 \\ \sqrt{64} & = & 8 & \text{because} & 8^2 & = & 64 \\ \sqrt{81} & = & 9 & \text{because} & 9^2 & = & 81 \\ \sqrt{100} & = & 10 & \text{because} & 10^2 & = & 100 \end{array}$$

## ESTIMATING OTHER SQUARE ROOTS

It's possible to take the square root of any positive number, but it can be tricky for numbers that are not perfect squares. For example, the square root of 2 is best represented as a decimal:

$$\sqrt{2} = 1.414213562....$$

(In fact, $\sqrt{2}$ is different from all of the numbers that we've seen up to this point. Its decimal expansion never repeats, and it cannot be represented as a whole-number fraction. Numbers such as $\sqrt{2}$ are called **irrational**–they can't be expressed as a ratio of two whole numbers.)

The important thing is that because $2$ is between $1$ and $4$, its square root is between $\sqrt{1}$ and $\sqrt{4}$ –that is, between $1$ and $2$. You can always estimate the square root of a number by sandwiching it between two perfect squares.

Let's see how this works. What can you say about $\sqrt{57}$?

Well, $57$ is between $49$ and $64$, so $\sqrt{57}$ must be between $\sqrt{49}$ and $\sqrt{64}$, or between $7$ and $8$. (And check it out with a calculator: $\sqrt{57} = 7.54983...$)

## SQUARING A SQUARE ROOT AND VICE VERSA

Taking the square root and squaring are two operations that undo each other. This means two things.

### Thing one

The square of the square root of any number is itself. Take a look.

What's $\left(\sqrt{36}\right)^2$?

Well, $36$ is a perfect square: $\sqrt{36} = 6$. And $6^2 = 36$.

So $\left(\sqrt{36}\right)^2 = 36$.

This is always true, even for nonperfect squares. So $\left(\sqrt{7}\right)^2 = 7$ and $\left(\sqrt{29}\right)^2 = 29$ and $\left(\sqrt{1729}\right)^2 = 1729$.

### Thing two

The square root of the square of any positive number is itself. Take another look.

What's $\sqrt{11^2}$?

Well, $11^2 = 121$, which means that $\sqrt{11^2} = \sqrt{121} = 11$. This is also always true. Try a few examples.

# MULTIPLYING AND DIVIDING SQUARE ROOTS

What do you get when you multiply $\sqrt{25}$ and $\sqrt{4}$?

Well $\sqrt{25}$ is $5$ and $\sqrt{4}$ is $2$, so $\sqrt{25} \times \sqrt{4} = 5 \times 2 = 10$. So far, so good.

But $25 \times 4 = 100$, which just happens to be the square of $10$.

Sure makes you think, doesn't it?

## THE MULTIPLICATION RULE

The product of two square roots is the square root of their product.

So $\sqrt{25} \times \sqrt{4} = \sqrt{100}$, which is, in fact, exactly what we got.

Let's try this one more time: what's $\sqrt{9} \times \sqrt{4}$?

If you do it directly, you get

$$\sqrt{9} \times \sqrt{4} = 3 \times 2 = 6.$$

If you use the rule, you get

$$\sqrt{9} \times \sqrt{4} = \sqrt{9 \times 4} = \sqrt{36} = 6.$$

Looks like the rule works: we got the same answer both times.

## THE MULTIPLICATION RULE WITH NONPERFECT SQUARES

So long as we're working with perfect squares, the rule looks like it just makes things more difficult. But take a look at this: what's $\sqrt{2} \times \sqrt{32}$?

Neither $2$ nor $32$ is a perfect square, so it's unclear how to do this problem directly. So use the multiplication rule:

$$\sqrt{2} \times \sqrt{32} = \sqrt{2 \times 32} = \sqrt{64}.$$

Luckily, $64$ *is* a perfect square. So $\sqrt{2} \times \sqrt{32} = \sqrt{64} = 8$: the multiplication rule made a difficult problem into an easy one!

## GOING THE OTHER WAY

The multiplication rule works in the other direction as well: the square root of a product is the product of the square roots. This can come in handy when you're working with bigger numbers.

For example, computing $\sqrt{49 \times 9}$ is a hassle. First you have to find $49 \times 9$; then you have to take the square root of the resulting three-digit number. But because both $49$ and $9$ are perfect squares, the computation can be simplified:

$$\sqrt{49 \times 9} = \sqrt{49} \times \sqrt{9} = 7 \times 3 = 21.$$

That's much nicer than figuring out the square root of $49 \times 9 = 441.$

## THE DIVISION RULE

As usual, there's an analogous rule for dividing square roots: the quotient of two square roots is the square root of their quotient.

Let's see how this works. What is $\frac{\sqrt{64}}{\sqrt{4}}$?

Well, $\sqrt{64}$ is $8$ and $\sqrt{4}$ is $2$, so

$$\frac{\sqrt{64}}{\sqrt{4}} = \frac{8}{2} = 4.$$

Now let's try using the division rule:

$$\frac{\sqrt{64}}{\sqrt{4}} = \sqrt{\frac{64}{4}} = \sqrt{16}.$$

And $\sqrt{16}$ is $4$, so we got the same answer! It looks like the rule works. Brilliant.

## THE DIVISION RULE WITH NONPERFECT SQUARES

Like the multiplication rule, the division rule is helpful when working with nonperfect-square numbers. Try this one:

What is $\frac{\sqrt{175}}{\sqrt{7}}$?

Tricky: neither $175$ nor $7$ is a perfect square. (How do I know this? I happen to know that $13^2 = 169$ and $14^2 = 196$, and $175$ is between these two.)

Use the division rule:

$$\frac{\sqrt{175}}{\sqrt{7}} = \sqrt{\frac{175}{7}} = \sqrt{25}.$$

Fortunately, $25$ is a perfect square, and $\frac{\sqrt{175}}{\sqrt{7}} = 5$.

### RULE SUMMARY

Here are both the root rules again, just so you don't forget.

**Multiplication rule**

$$\sqrt{\text{multiplicand}} \times \sqrt{\text{multiplier}} = \sqrt{\text{multiplicand} \times \text{multiplier}}$$

**Division rule**

$$\frac{\sqrt{\text{numerator}}}{\sqrt{\text{denominator}}} = \sqrt{\frac{\text{numerator}}{\text{denominator}}}$$

## SIMPLIFYING SQUARE ROOTS

Say you're dealing with a $\sqrt{50}$. Since $50 = 25 \times 2$, you can rewrite:

$$\sqrt{50} = \sqrt{25 \times 2} = \sqrt{25} \times \sqrt{2}.$$

And since $\sqrt{25} = 5$, you know that

$$\sqrt{50} = \sqrt{25} \times \sqrt{2} = 5 \times \sqrt{2}.$$

By the way, people usually drop the multiplication sign and write $5\sqrt{2}$ instead of $5 \times \sqrt{2}$. There's no confusion because the radical sign breaks things up; it's like writing $7(3 + 4)$ instead of $7 \times (3 + 4)$.

So $\sqrt{50}$ is the same as $5\sqrt{2}$. Which one is better?

Neither, really. Both represent the exact same quantity: about $7.071$ (makes sense, since $50$ is close to $49 = 7^2$). However, $5\sqrt{2}$ is formally considered a more simplified form than $\sqrt{50}$, because the factor of $25$, which is a perfect square, is gone from under the radical.

The emphasis on simplifying expressions with square roots is dying out as more and more people use calculators all the time. But some teachers still care about it, so I want to go over it. If yours happens not to, feel free to skim this section lightly.

## SIMPLIFIED FORM

A square root expression is considered **simplified** if the radicand (the bit under the radical sign) has no repeated factors.

For example, $\sqrt{30}$ is in simplified form, because $30 = 2 \times 3 \times 5$; all three factors are different. On the other hand, $\sqrt{40}$ can be simplified further: $40 = 2 \times 2 \times 2 \times 5$; since $2$ appears three times, it can be pulled out from under the radical:

$$\sqrt{40} = \sqrt{4 \times 10} = \sqrt{4} \times \sqrt{10} = 2\sqrt{10}.$$

Actually, it's enough to check for repeated *prime* factors: if a factor appears more than once, then the primes that it's divisible by appear more than once.

## SIMPLIFYING STEP BY STEP

To simplify a square root:

1. **Factor** the radicand into its prime factors. (Check back to Chapter 6 to see how to find the prime factorization of a number.)
2. If any prime factors are repeated, **group** them into pairs.
3. **Pull out** each pair from under the radical sign. (Watch out: a factor of $3^2$ under the radical sign means a factor of $3$ outside.)
4. **Simplify** by multiplying together all the factors under the radical sign and multiplying together all the factors outside.

Let's do an example to see how this works.

Express $\sqrt{756}$ in simplified form:

1. **Factor:** $756 = 2 \times 2 \times 3 \times 3 \times 3 \times 7$.

$$\sqrt{756} = \sqrt{2 \times 2 \times 3 \times 3 \times 3 \times 7}$$

2. **Regroup** to make pairs: There are two $2$s and three $3$s in the prime factorization of $756$, so the two $2$s form a pair and two of the $3$s form a pair. The third $3$ is left out in the cold.

$$\sqrt{2^2 \times 3^2 \times 3 \times 7}$$

3. **Pull out** repeated factors: Technically, we're using the multiplication rule here, but pretty soon you should be able to skip the intermediate steps.

$$\sqrt{2^2 \times 3^2 \times 3 \times 7} = \sqrt{2^2} \times \sqrt{3^2} \times \sqrt{3 \times 7}$$

$$= 2 \times 3 \times \sqrt{3 \times 7}$$

4. **Simplify** by multiplying factors together.

$$2 \times 3 \times \sqrt{3 \times 7} = 6\sqrt{21}$$

So $\sqrt{756} = 6\sqrt{21}$ Perfect.

## SIMPLIFYING FREE-FOR-ALL

Actually, if you're good at eyeballing factors, you can just start pulling them out. It's just like simplifying fractions: you can either find prime factorizations and cancel factors methodically, or you can just start canceling whatever you happen to notice. Either way you get the same final result.

For example, if you noticed that $756$ is divisible by $4 = 2^2$, you can pull out a factor of $2$ right away:

$$\sqrt{756} = \sqrt{4 \times 189} = 2\sqrt{189}.$$

And if you then notice that $189$ is divisible by $9 = 3^2$, you can pull out a factor of $3$:

$$2\sqrt{189} = 2\sqrt{9 \times 21} = 2 \times 3\sqrt{21} = 6\sqrt{21}.$$

SQUARE ROOTS

At this point, since you know that $21$ has no repeated factors ($21 = 3 \times 7$), you're done. Isn't it nice that you got the same answer both ways?

## MORE SIMPLIFYING

Simplifying works exactly the same way if you're starting out with a product of a number and a square root. Take a look at this example.

Simplify $\frac{3\sqrt{800}}{5}$.

1. **Factor.**

$$\frac{3\sqrt{800}}{5} = \frac{3\sqrt{2 \times 2 \times 2 \times 2 \times 2 \times 5 \times 5}}{5}$$

2. **Regroup:** There are five factors of $2$ and two factors of $5$. Four $2$s make two pairs; the fifth is left over.

$$\frac{3\sqrt{2^2 \times 2^2 \times 5^2 \times 2}}{5}$$

3. **Pull out.**

$$\frac{3 \times 2 \times 2 \times 5\sqrt{2}}{5}$$

4. **Simplify.**

$$\frac{3 \times 2 \times 2 \times \cancel{5}\sqrt{2}}{\cancel{5}} = 12\sqrt{2}$$

So $\frac{3\sqrt{800}}{5} = 12\sqrt{2}$.

Last thing to notice: $\sqrt{2 \times 2 \times 2 \times 2}$ is the same thing as $\sqrt{2^4}$, or $2^2$: the square root of an even power can be simplified by dividing the exponent in half:

$$\sqrt{\text{number}^{\text{even power}}} = \text{number}^{(\text{even power}) \div 2}.$$

## SQUARE ROOT IN THE DENOMINATOR

Any expression with a root in the denominator is considered not simplified. To get the square root out of there, we can do a nifty little trick called **rationalizing the denominator**.

Say we're dealing with $\frac{2}{\sqrt{3}}$. To derootify the denominator, multiply the fraction by $\frac{\sqrt{3}}{\sqrt{3}}$. That's just a special form of $1$, so it doesn't change the value of the fraction:

$$\frac{2}{\sqrt{3}} \times \frac{\sqrt{3}}{\sqrt{3}} = \frac{2 \times \sqrt{3}}{\sqrt{3} \times \sqrt{3}}.$$

The $\sqrt{3} \times \sqrt{3}$ in the denominator is the same thing as $\left(\sqrt{3}\right)^2$, or $3$. (Alternatively, you can think about it as $\sqrt{3 \times 3} = \sqrt{9}$.) So the fraction becomes

$$\frac{2}{\sqrt{3}} = \frac{2}{\sqrt{3}} \times \frac{\sqrt{3}}{\sqrt{3}} = \frac{2\sqrt{3}}{3}.$$

Neat, isn't it? The $\sqrt{3}$ has traveled out of the denominator into the numerator, leaving a factor of $3$ in its wake. This is one of my favorite tricks. (Incidentally, it's called *rationalizing* the denominator because you're making the denominator into a so-called *rational* number, one that doesn't involve any square roots.)

Let's do a couple more examples to make sure that you've got the hang of it.

Rationalize the denominator of $\frac{7\sqrt{3}}{2\sqrt{21}}$.

Again, multiply top and bottom by the square root in the denominator—that is, $\sqrt{21}$:

$$\frac{7\sqrt{3}}{2\sqrt{21}} \times \frac{\sqrt{21}}{\sqrt{21}} = \frac{7\sqrt{3} \times \sqrt{21}}{2\sqrt{21} \times \sqrt{21}}.$$

To clean up the numerator, mash all the square root factors together and simplify the square root:

$$\frac{7\sqrt{3} \times \sqrt{21}}{2\sqrt{21} \times \sqrt{21}} = \frac{7\sqrt{3 \times 21}}{2 \times 21} = \frac{7\sqrt{3 \times 3 \times 7}}{42} = \frac{7 \times 3\sqrt{7}}{42}.$$

SQUARE ROOTS

Finally, reduce the fraction:

$$\frac{\cancel{7} \times \cancel{3}\sqrt{7}}{\cancel{42}_{2}} = \frac{\sqrt{7}}{2}.$$

So $\dfrac{7\sqrt{3}}{2\sqrt{21}} = \dfrac{\sqrt{7}}{2}$.

If you're feeling enterprising, you can often save yourself some of this busywork. The fraction $\dfrac{7\sqrt{3}}{2\sqrt{21}}$ has a factor of $\sqrt{3}$ both in the numerator and in the denominator (it's hiding in $\sqrt{21}$). So you can reduce first, then rationalize. Like so:

**Reduce:**

$$\frac{7\sqrt{3}}{2\sqrt{21}} = \frac{7\cancel{\sqrt{3}}}{2\cancel{\sqrt{21}}_{\sqrt{7}}} = \frac{7}{2\sqrt{7}}$$

**Rationalize:**

$$\frac{7}{2\sqrt{7}} \times \frac{\sqrt{7}}{\sqrt{7}} = \frac{7\sqrt{7}}{2\sqrt{7} \times \sqrt{7}} = \frac{7\sqrt{7}}{2 \times 7}$$

**Finally, reduce again:**

$$\frac{\cancel{7}\sqrt{7}}{2 \times \cancel{7}} = \frac{\sqrt{7}}{2}$$

The answer is the same, and the numbers that you're dealing with are a little smaller, which is nice. But the bottom line is this: you know how to reduce fractions, you know how to multiply square roots, and you know how to rationalize denominators. So as long as every step you're doing is legitimate, the order is up to you.

# ADDING AND SUBTRACTING SQUARE ROOTS

Have you noticed that we've talked a ton about multiplying square roots but not much about adding them? That's because there isn't really anything helpful to know. Not much you can do with

$$\sqrt{11} + \sqrt{3} \quad \text{or with} \quad \sqrt{30} - \sqrt{5}$$

except grin and bear it.

## "LIKE" TERMS

The one thing you *can* do, however, is rewrite $\sqrt{7} + 2\sqrt{7}$ as $3\sqrt{7}$. You can think of it like adding one orange and two oranges to make three—the $\sqrt{7}$ s are the oranges. (A blast from the past: technically, this works because of the distributive property: $1\sqrt{7} + 2\sqrt{7} = (1+2)\sqrt{7}$.)

This kind of adding oranges and oranges is often called **combining "like" terms**—the $\sqrt{7}$ and the $2\sqrt{7}$ are *like* each other because both terms are numbers multiplied by a $\sqrt{7}$.

Try simplifying this:
$8 - \sqrt{3} + 6\sqrt{5} - 3 - 2\sqrt{5} + 4\sqrt{3}$.

There are three kinds of terms here: ordinary numbers, $\sqrt{3}$ -like terms, and $\sqrt{5}$ -like terms. Deal with each kind separately, but be super-careful about the signs.

Ordinary numbers: $8 - 3 = 5$
$\sqrt{3}$ -like terms: $4\sqrt{3} - \sqrt{3} = 3\sqrt{3}$
$\sqrt{5}$ -like terms: $6\sqrt{5} - 2\sqrt{5} = 4\sqrt{5}$

So $8 - \sqrt{3} + 6\sqrt{5} - 3 - 2\sqrt{5} + 4\sqrt{3} = 5 + 3\sqrt{3} + 4\sqrt{5}$.

## SIMPLIFY, THEN COMBINE

It's usually hard to tell which square root terms are like which others until you simplify—and that includes rationalizing the denominator.

Try this one out: $\sqrt{45} - \frac{4}{\sqrt{20}} + \frac{\sqrt{10}}{\sqrt{5}}$.

Can you guess which two of these terms can be combined? Let's see if you guessed right by simplifying, term by term.

$\sqrt{45}$ : This one looks like it might have a repeated factor under the square root: $45$ is divisible by $9$, which is a perfect square.

$$\sqrt{45} = \sqrt{3^2 \times 5} = 3\sqrt{5}$$

$\frac{4}{\sqrt{20}}$ : This term needs some denominator rationalization. But before we do it, let's make sure that the square root term is as simple as it can be.

$$\frac{4}{\sqrt{20}} = \frac{4}{\sqrt{2^2 \times 5}} = \frac{4}{2\sqrt{5}} = \frac{2}{\sqrt{5}}$$

Now rationalize:

$$\frac{2}{\sqrt{5}} = \frac{2}{\sqrt{5}} \times \frac{\sqrt{5}}{\sqrt{5}} = \frac{2\sqrt{5}}{5}$$

$\frac{\sqrt{10}}{\sqrt{5}}$ : This term has a factor of $\sqrt{5}$ both in the numerator (where it's hidden inside $\sqrt{10}$) and in the denominator. Clean it up!

$$\frac{\sqrt{10}}{\sqrt{5}} = \frac{\cancel{\sqrt{5}} \times \sqrt{2}}{\cancel{\sqrt{5}}} = \sqrt{2}$$

After reducing each term, we're left with $3\sqrt{5} - \frac{2\sqrt{5}}{5} + \sqrt{2}$. It looks like the first and second terms can be combined. Since $3 - \frac{2}{5} = \frac{13}{5}$, the final result is

$$\frac{13\sqrt{5}}{5} + \sqrt{2}.$$

Did you guess right?

# CUBE ROOTS AND OTHERS

If taking the square root undoes squaring, then what kind of operation undoes cubing? And what kind of operation undoes raising to fourth or fifth powers?

These operations exist, all right. For example, the **cube root** of a number is the number that, when cubed, will give the original. So the cube root of $8$ is $2$, because $2^3 = 8$.

The cube root sign is $\sqrt[3]{\ }$; it looks just like the square root sign with a little $3$ attached:

$$\sqrt[3]{8} = 2 \quad \text{because} \quad 2^3 = 8$$

$$\sqrt[3]{27} = 3 \quad \text{because} \quad 3^3 = 27$$

$$\sqrt[3]{64} = 4 \quad \text{because} \quad 4^3 = 64$$

You can go on and on. For example, $\sqrt[3]{1{,}000{,}000} = 100.$

Like square roots, cube roots are well behaved under multiplication and division. For example,

$$3\sqrt{8 \times 125} = \sqrt[3]{8} \times \sqrt[3]{125} = 2 \times 5.$$

The **fourth root** of a number is the number that, when raised to the fourth power, gives the original number:

$$\sqrt[4]{16} = 2 \quad \text{because} \quad 2^4 = 16.$$

And you can go and on, taking fifth and sixth and seventh and tenth roots. And all of these have a multiplication rule and a division rule. You can even multiply them together and rationalize their denominators. But all of that is for a different book. Lucky you, eh?

# YOUR TURN

Solutions start on page 301.

1. Find each of these square roots.

   (a) $\sqrt{1,000,000}$

   (b) $\sqrt{279}$

   (c) $\sqrt{576}$

   (d) $\sqrt{49 \times 49}$

   (e) $\sqrt{\frac{64}{9}}$

   (f) $\sqrt{5^6}$

   (g) $\sqrt{3^5 \times 3^3}$

   (h) $\sqrt{6.76}$

   (i) $\sqrt{2^5 - 2^4}$



2. A *Pythagorean triple* is three whole numbers with the property that the sum of the squares of the first two numbers equals the square of the third. For example, $3$, $4$, and $5$ are a Pythagorean triple because $3^2 + 4^2 = 5^2$. (Check it: $9 + 16 = 25$.)

   Complete these Pythagorean triples.

   (a) $6^2 + 8^2 = \_\_^2$

   (b) $8^2 + 15^2 = \_\_^2$

   (c) $5^2 + 12^2 = \_\_^2$

   (d) $7^2 + \_\_^2 = 25^2$

3. Simplify.

   (a) $3\sqrt{25}$

(b) $\sqrt{72}$

(c) $\sqrt{115}$

(d) $\sqrt{252}$

(e) $\sqrt{128}$

(f) $\sqrt{2^6 \times 3^9 \times 5^3 \times 7^2}$

4. Simplify.

(a) $6\sqrt{2} \div 2\sqrt{18}$

(b) $\dfrac{1}{\sqrt{7}}$

(c) $\dfrac{21}{\sqrt{98}}$

(d) $\dfrac{4\sqrt{24}}{\sqrt{32}}$

(e) $\sqrt{27} + \sqrt{9} - \dfrac{4}{\sqrt{3}}$

5. Find these higher-power roots.

(a) $\sqrt[3]{1{,}000{,}000}$

(b) $\sqrt[3]{216}$

(c) $\sqrt[3]{1728}$

(d) $\sqrt[4]{256}$

(e) $\sqrt[10]{1024}$

(f) $\sqrt[6]{10^{12}}$

# 14

# CHAPTER 14
# POSITIVE AND NEGATIVE NUMBERS

## ROCK 'n' ROLL HIGH SCHOOL

"So we're all in agreement?" Dad's beautiful blond secretary asked, looking around the cramped Saturn. She was jammed between Kyle and Jake in the backseat. Everyone nodded.

"Excellent!" I said. "As soon as Johnny London debuts on Pfufferfish and skyrockets up the charts, peaceful rock 'n' roll charities will become flush with cash."

"Can I go now?" Dad's beautiful secretary asked. "I have a babka baking in the drippy cave's oven, and if it overcooks even by a minute . . ."

"Say no more," Kyle said, opening his door and stepping out. Everyone got out and shook hands, and Dad's beautiful blond secretary pedaled off back to the cave to tell my parents and the Coast Guard reserves that serious cultural and structural upheaval had been successfully, if narrowly, avoided.

Before heading back, we all followed Kyle to the nearby scenic overlook, a vista of power lines and maple trees. Jake put his arm around Rebecca's shoulders.

Wait. Jake put his arm around Rebecca's shoulders!

"What's going on here?" I asked.

"Oh yeah, I forgot that you don't know everything yet," Rebecca said. "Jake and I are an item. We've been going out for over a year now. When we came to Maplewood as rogue spies, we had to keep our relationship a secret. But now that our spying days are behind us, we can finally express our love for each other in public."

And with that, Rebecca gave Jake a kiss!!! Which left Kyle and me just kind of standing there, examining our shoes, checking our watches, peering out at the scenic power lines—looking anywhere but into each other's eyes.

"So are you, like, just going to quit high school now?" I asked Kyle, staring at a bug on the railing.

"Oh, no!" he said, kicking at some dirt clods. "I did that back in England, mate. Big mistake. This time, I'm going to stick with it and graduate. It's going to be tricky with Johnny London taking up so much of my time, but I think I really need to get a diploma."

"Oh, great!" I said. ". . . You know, there's a math quiz tomorrow."

"I know," he said. "On multiplying negative numbers! That's, like, even worse than rationalizing denominators. I could really use some help."

"Finer Diner?" I offered.

"Finer with me!" he laughed.

* * * * *

If we subtract $3$ from $7$, we get $4$. On the number line, we start at $7$ and move $3$ units to the left.

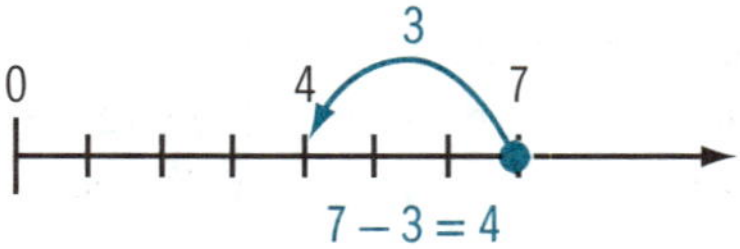


But what happens if we try to subtract $7$ from $3$? What's going on on the dark side of zero?

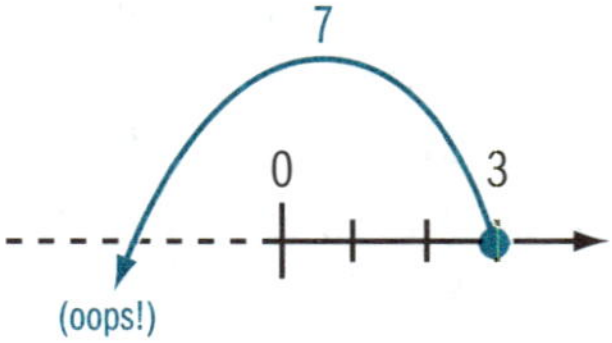

# POSITIVE AND NEGATIVE NUMBERS

You already know everything about **positive numbers**. These are all the numbers that we've been playing around with so far—natural numbers, fractions, square roots, decimals. In fact, the only number we've talked about that isn't positive is $0$.

On the number line, positive numbers are represented by all the points to the right of $0$.

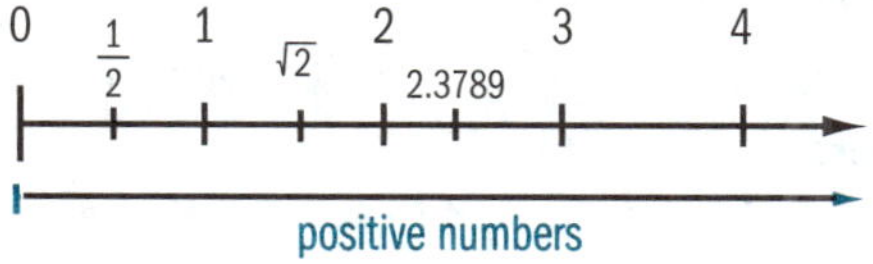


## NEGATIVE NUMBERS

Many real-world quantities—apples eaten, distance traveled, days gone by—are measured with positive numbers. But sometimes it turns out to be useful to measure things in two directions: Both distance traveled north and distance south. Both seconds after takeoff and seconds before takeoff. Both floors above ground and floors below. Both apples eaten and apples thrown up.

That's where negative numbers come in. If you pick a neutral starting point and call it zero, then you can count both positive numbers in one direction and negative numbers in the other direction.

On the number line, negative numbers are all the points to the left of $0$.

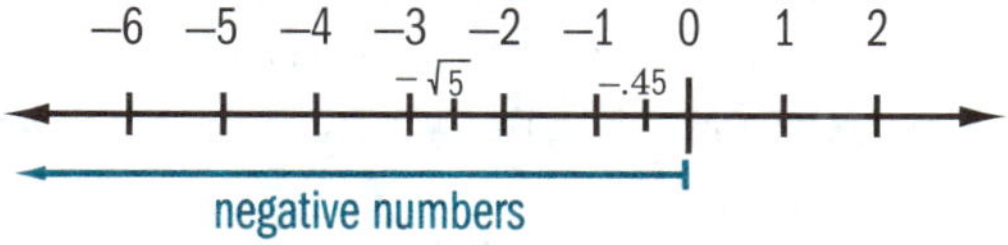


The point $-3$ is the number three *less than zero*. It's identified as "negative three" or, less formally, as "minus three."

Some people make a stink about the difference between the **minus sign**, which marks subtraction in expressions such as $11 - 3$, and the **negative sign**,

POSITIVES & NEGATIVES

which is typeset slightly raised and can introduce a negative number, like this: $^{-}3$. These silly sticklers say that the minus sign is an operation on two numbers, whereas a negative sign identifies a single number, yadda, yadda, yadda. I don't really think it's worth arguing about, but down deep, the negative and the minus are really the same kind of subtracting-lessening-minusing-negating creature. As you'll see in a minute, adding a negative number and subtracting a positive number is the same thing. And $-3$ is exactly what you get when you compute $0 - 3$; you can just think of it as dropping the $0$.

## THE REAL NUMBER LINE

The collective name for all the numbers represented by points on the number line is **real numbers**.

Every real number except zero is either positive or negative. Zero is neither positive nor negative; it's in a class by itself.

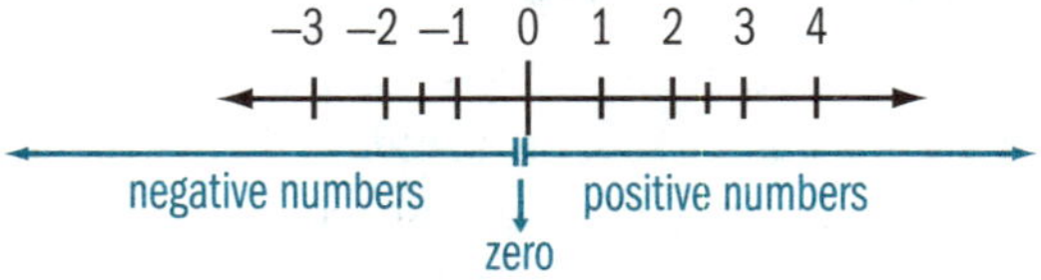


The natural numbers, their negatives, and zero, all together $(\ldots, -3, -2, -1, 0, 1, 2, 3, \ldots)$, are called **integers**.

## ABSOLUTE VALUE

The **absolute value** of a number is its distance from $0$. For example, the number $4$ is four units away from $0$; its absolute value is $4$. The number $-4$ is also four units away from $0$, in the other direction; its absolute value is also $4$.

Taking the absolute value is marked with two vertical lines, on either side:

$$|-4| = 4.$$

This is read as "the absolute value of negative four is four."

Absolute values are always nonnegative—either positive or zero.

A few examples to get the absolute-value juices flowing:

$$\begin{aligned} |-1000| &= 1000 \\ |1978| &= 1978 \end{aligned}$$

$$|0| = 0$$

$$\left|\frac{7}{2}\right| = \frac{7}{2}$$

$$\left|-\frac{7}{2}\right| = \frac{7}{2}$$

$$|-0.6768| = 0.6768$$

$$|-\sqrt{2}| = \sqrt{2}$$

In math expressions, absolute value marks behave like parentheses. Evaluate the quantity inside before venturing outside.

Try out this example:

$$2\left|4 + 3^2\right| - |-7|$$

Figure out $\left|4 + 3^2\right|$ and $|-7|$ before moving on:

$\left|4 + 3^2\right| = |4 + 9| = |13| = 13$, and $|-7| = 7$.

So the expression becomes

$$2 \times 13 - 7,$$

which evaluates to $19$.

## PLUSES, MINUSES, AND ZERO

For symmetry's sake, since negative numbers are introduced by minus signs, positive numbers can be introduced by plus signs. So $16$ can also be written as $+16$. But people usually don't bother writing that plus sign except in special circumstances, when they're drawing a distinction. It's the same in speech: you can say "positive eleven" if you're contrasting with "negative eleven," but most of the time you just say "eleven."

Every number has an opposite—that's the number with the same absolute value and the opposite sign. The opposite of $+3$ is $-3$; the opposite of $-22$ is $+22$; the opposite of $-\frac{2}{5}$ is $\frac{2}{5}$.

In fact, you can actually think about the minus sign as the introducer of *opposites* rather than negatives. So $-3$ means "the opposite of $3$," which is just negative $3$. The opposite of $-3$, or rather, "the opposite of the opposite of $3$," is written as $-(-3)$ and is equal to $+3$. Two opposites cancel each other out.

Every number on the number line can be paired with its opposite. (The only exception is 0; more on that in a moment.) So $1$ and $-1$ form a pair, and $-12.75$ and $12.75$ form a pair. The two numbers in such a pair are the same distance away from zero, just in different directions. They have the same absolute value.

So what's up with 0? It's neither positive nor negative. It's neutral; it can't be affected by signs. If you like, you can write +0 instead of 0. And 0 is the only number that is its own opposite, so –0 is just 0 again. These three ways of writing (0, +0, and –0) all indicate the same point on the number line.

## COMPARING SIGNED NUMBERS

On the number line, comparing signed numbers is just like comparing positive numbers: the number farther to the right is always greater.

For example, 0 is greater than $-7$. And $-15$ is greater than $-100$.

If you're comparing . . .

- **Two positive numbers:** The one with the bigger absolute value is greater.

  $10{,}000$ is greater than $4$

- **A positive number and a negative number:** The positive number is greater.

  $302$ is greater than $-2$
  $2$ is greater than $-302$

- **Two negative numbers:** The one with the *smaller* absolute value is greater.

  $-5$ is greater than $-2$
  $-7.5$ is greater than $-100$

- **Zero and a positive number:** The positive number is greater.

  $34$ is greater than $0$

- **Zero and a negative number:** Zero is greater.

  $0$ is greater than $-57$

POSITIVES & NEGATIVES

# ADDING AND SUBTRACTING SIGNED NUMBERS

Let's go back to that number line for a sec. If you're working with positive numbers, adding means moving to the right:

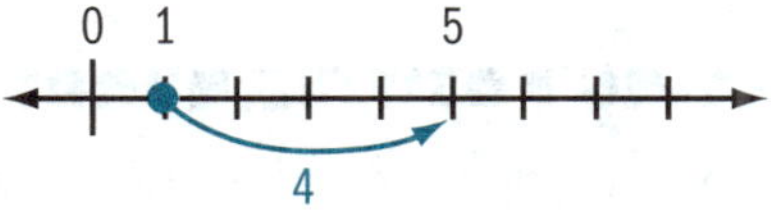


$1 + 4 = 5$

Subtracting means moving to the left:

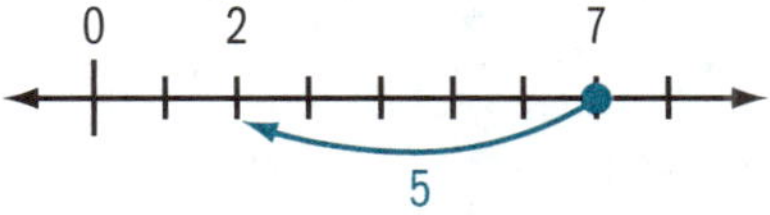


$7 - 5 = 2$

So far, so good. Fortunately, the same thing happens even when you start out on the other side of zero.

## ADDING POSITIVE NUMBERS

When adding a positive number, move to the right. See what happens when you add $4$ to $-7$:

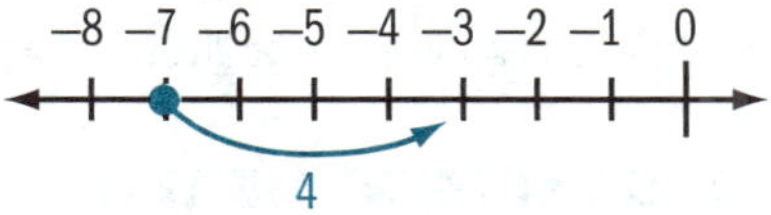


Start at $-7$ and move $4$ steps to the right, to land at $-3$.

So $-7 + 4 = -3$.

Don't be perturbed if you have to pass $0$. Just keep on going. For example, take a look at $-3 + 5$:

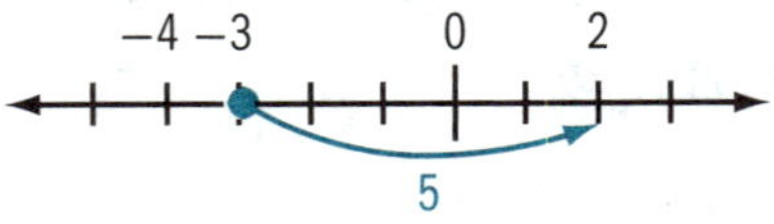


So $-3 + 5 = 2$.

## SUBTRACTING POSITIVE NUMBERS

When subtracting a positive number, move to the left. Here's what happens with $-4 - 2$:

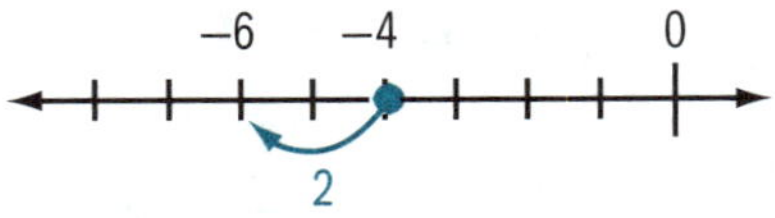


So $-4 - 2 = -6$.

Again, plunge right ahead even if you stumble upon $0$. Here, for example, is $3 - 6$:

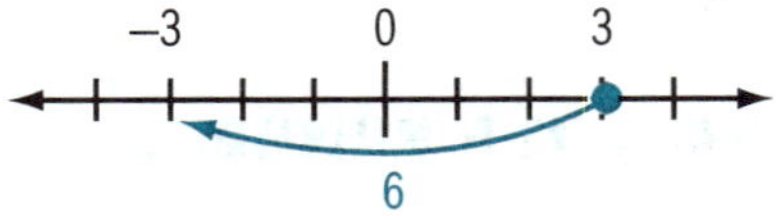


So $3 - 6 = -3$.

Now that you can venture into the negatives, you can subtract any number from any other—the first no longer needs to be greater than the second.

## ADDING NEGATIVE NUMBERS

Here's where things get a little different. If you're adding or subtracting a negative number, move *the other way*. So to add a negative number, move left. To subtract a negative number, move right.

For example, here's $5 + (-2)$:

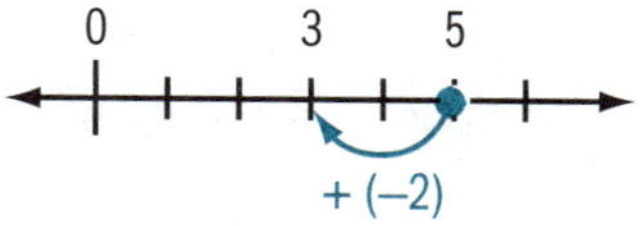


So $5 + (-2) = 3$.

And here's $3 + (-8)$:

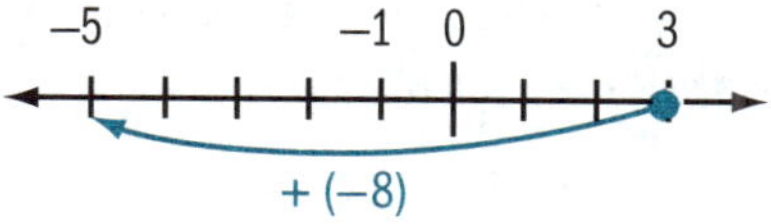


So $3 + (-8) = -5$.

### Adding negatives, Subtracting positives

Hmm, you might say. When I add negative numbers, I move left. When I subtract positive numbers, I move left. Suspicious.

Suspicious indeed. Adding a negative number *is exactly the same thing as* subtracting the corresponding positive number. Let me say that again:

> **Adding a negative number is the same thing as subtracting the opposite positive number.**

So $3 + (-8)$ is the same thing as $3 - 8$. And $5 + (-2)$ is the same thing as $5 - 2$.

Take a look at another number line:

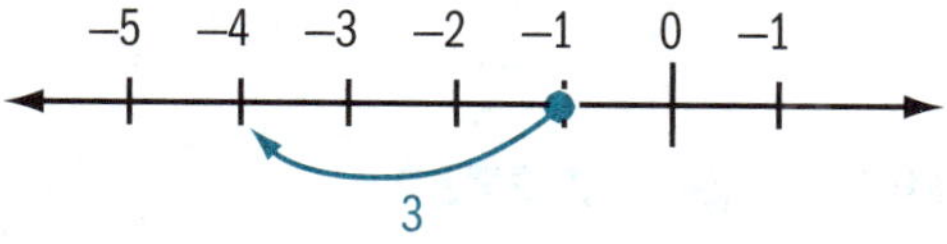


We start at $-1$ and move $3$ to the left to get to $-4$. This can be interpreted as

$$-1 - 3 = -4.$$

On the other hand, this can also be interpreted as

$$-1 + (-3) = -4.$$

The two equations are equivalent; they're expressing the same thing in slightly different ways.

## SUBTRACTING NEGATIVE NUMBERS

As I mentioned, when you're adding or subtracting a negative number, you have to move the other way. So to subtract a negative number, move right.

Here's the example of $-1 - (-4)$:

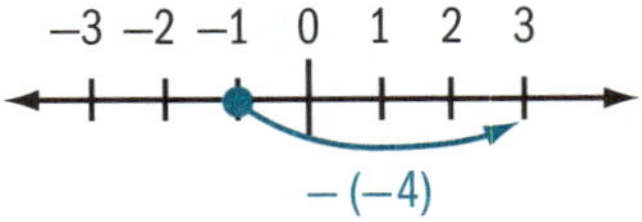


So $-1 - (-4) = 3$.

And this is the same as computing $-1 + 4 = 3$. The two negative signs before the $4$ make a positive.

### Subtracting negatives, Adding positives

Subtracting a negative means moving to the right. So does adding a positive. This means that

> **Subtracting a negative number is the same thing as adding the opposite positive number.**

See if you can figure out how this works without the number line: What's $3 - (-2)$?

Since subtracting $-2$ is the same thing as adding $2$, we can rewrite:

$$3 - (-2) = 3 + 2 = 5.$$

## ADDING STEP BY STEP

If you have a good feeling for how positive and negative numbers work, feel free to skip the next two sections. But if you prefer a method, here we go.

**To add two same-sign numbers:**

1. **Add** their absolute values to get the absolute value of the sum.
2. The sign of the sum is the sign of either number.

So to add $4$ and $5$, add their absolute values to get $9$ and give the sum a positive sign. You already know what this looks like:

$$4 + 5 = 9.$$

To add $-4$ and $-5$, add their absolute values to get $9$ and give the sum a negative sign:

$$-4 + -5 = -9.$$

**To add two different-sign numbers:**

**1. Subtract** their absolute values to get the absolute value of the sum.
**2.** The sign of the sum is the sign of the number with the larger absolute value.

So to add $6$ and $-2$, subtract $2$ from $6$ to get $4$. Since $6$ is greater than $2$, the sign of the sum is positive:

$$6 + (-2) = 4.$$

To add $-6$ and $2$, subtract $2$ from $6$ to get $4$. Since $6$ is greater than $2$, the sign of the sum is negative:

$$-6 + 2 = -4.$$

## SUBTRACTING STEP BY STEP

Instead of learning new rules, we use the fact that subtracting is just adding the opposite.

**To subtract signed numbers:**

**1. Change** the subtraction problem into an addition problem by reversing the sign of the second number.
**2.** Add using the rules above.

We'll do four examples, rapid-fire.

**First example:** $-7 - (-5)$

**1.** Convert to $-7 + 5$.
**2.** The numbers have different signs, so the sum will have absolute value $2$. Since $7$ is greater than $5$, the sum is negative:

So $-7 - (-5) = -7 + 5 = -2$.

**Second example:** $5 - (-9)$

1. Convert to $5 + 9$.
2. Add to get $14$.

So $5 - (-9) = 5 + 9 = 14$.

**Third example:** $4 - 11$

1. Convert to $4 + (-11)$.
2. The numbers have different signs, so the sum will have absolute value $7$. Since $11$ is greater than $4$, the sum is negative.

So $4 - 11 = 4 + (-11) = -7$.

**Fourth example:** $-2 - 8$

1. Convert to $-2 + (-8)$
2. The numbers have the same sign, so the sum will have absolute value $10$. Moreover, the sum is negative.

So $-2 - 8 = -2 + (-8) = -10$.

### IS ADDITION STILL COMMUTATIVE?

You bet. Remember that *commutative* just means that the order doesn't matter. And the addition rules don't talk about which number comes first; they only care about whether they have the same sign and what sign the larger-valued number has. So $-4 + 9$ and $9 + (-4)$ are evaluated the same way: since they have different signs, the absolute value of the answer is $5$. Since $9$ is greater than $4$, the answer is positive:

$$-4 + 9 = 9 + (-4) = 5.$$

## MULTIPLYING AND DIVIDING SIGNED NUMBERS

Multiplication and division with signed numbers is a breeze. Forget about the signs and multiply or divide normally. Then pick the right sign. Some examples to get you started:

$$12 \times (-4) = -48$$
$$(-60) \div (-5) = 12$$

$$(-2)(-11) = 22$$

$$\frac{-100}{25} = -4$$

## TWO POSITIVE NUMBERS

A product or quotient of two positive numbers is positive. But you already knew this:

$$4 \times 5 = 20$$

$$36 \div 3 = 12$$

## ONE POSITIVE, ONE NEGATIVE

A product or quotient of a positive number and a negative number is negative. It doesn't matter which—positive or negative—comes first.

$$(-8) \times 9 = -72$$

$$7 \times (-6) = -42$$

$$(-54) \div 9 = -6$$

$$56 \div (-7) = -8$$

## TWO NEGATIVES

A product or quotient of two negative numbers is positive. This is similar to what we were talking about before about two opposites canceling each other out.

$$(-4) \times (-15) = 60$$

$$(-84) \div (-6) = 14$$

## SIGN RULE SUMMARY

Here's all this sign information gathered together. It never hurts to go over things twice, you know.

**Multiplication**

$$(+) \times (+) = (+)$$

$$(+) \times (-) = (-)$$

$$(-) \times (+) = (-)$$

$$(-) \times (-) = (+)$$

**Division**

$$(+) \div (+) = (+)$$
$$(+) \div (-) = (-)$$
$$(-) \div (+) = (-)$$
$$(-) \div (-) = (+)$$

In other words, multiplying or dividing by a positive number keeps the sign the same; multiplying or dividing by a negative number flips the sign.

As usual, multiplying any number, positive or negative, by 0 gives 0. Dividing 0 by any number also gives 0. And dividing by 0 is still not allowed.

## MULTIPLYING BY –1

Nothing new here, but it's something worth pointing out. To multiply by –1, just flip the sign (that's because multiplying by 1 leaves the number unchanged):

$$4 \times -1 = -4$$
$$-6 \times -1 = 6$$
$$-1 \times -1 = 1$$

So multiplying by –1 gives you the opposite of the number you started with. This works for expressions, too. For example,

$$(2 + 5) \times (-1) = -(2 + 5),$$

which may help to explain . . .

## DISTRIBUTING THE MINUS SIGN REVISITED

Remember that funny rule about distributing minus signs by flipping all the signs inside parentheses? We went over it in Chapter 5:

$$40 - (4 + 3) \quad = \quad 40 - 4 - 3.$$

Well, now we have enough ammunition to figure out why it works. The minus sign in front of $(4 + 3)$ can be thought of as a multiplication by –1:

$$40 - (4 + 3) \quad = \quad 40 + (-1) \times (4 + 3).$$

Now, the regular distributive property that allows us to distribute the –1 over the sum in parentheses:

$$40 + (-1) \times 4 + (-1) \times 3.$$

And the rules of multiplying by negative numbers let us simplify the individual products:

$$40 + (-4) + (-3).$$

Finally, the rules of adding negative numbers let us get rid of the parentheses:

$$40 - 4 - 3.$$

So all of this manipulation gets us back to where we started:

$$40 - (4 + 3) = 40 - 4 - 3.$$

Distributing the minus sign is shorthand for distributing a $-1$. Aren't shorthands nice?

Let's do one more example of distributing the minus sign. Remember, flip every sign inside the parenthesis expression:

$$-(12 + 34 - 8 + 9) - (-22 - 7).$$

We can distribute the first minus sign to get

$$-12 - 34 + 8 - 9 - (-22 - 7),$$

and the second minus sign to get

$$-12 - 34 + 8 - 9 + 22 + 7.$$

# SIGNED FRACTIONS

Good news: signed fractions follow all the same rules that signed integers do. The only reason I brought them up at all is to point out a minor fact about the minus sign in a negative fraction.

It can go on **top**: $\frac{-1}{2}$. This means that you're thinking about $(-1) \div 2$.

It can go on the **bottom**: $\frac{1}{-2}$. This means that you're thinking about $1 \div (-2)$.

It can go in **front**: $-\frac{1}{2}$. This means that you're thinking about the opposite of $\frac{1}{2}$.

All of these represent the same quantity, a number halfway between $-1$ and $0$, also sometimes identified as $-0.5$.

That's all I want to say about signed fractions. Well, almost. I worked out a few examples for Kyle. See if you get the same answers I did.

$$\left(-\frac{1}{6}\right) + \left(-\frac{1}{3}\right) = -\frac{1}{2}$$

$$-\left(\frac{2}{5} - \frac{1}{2}\right) = \frac{1}{10}$$

$$\frac{-5}{8} \times \frac{6}{-25} = \frac{3}{20}$$

$$\left(-\frac{5}{7}\right) \div \frac{5}{8} = -\frac{8}{7}$$

POSITIVES & NEGATIVES

# YOUR TURN

Solutions start on page 305.

1. For each pair, determine whether the two numbers represent the same quantity or different quantities.

(a) $|-13|$ and $-13$

(b) $|4.5|$ and $4.5$

(c) $\left|-\frac{3}{4}\right|$ and $\frac{3}{4}$

(d) $-|67|$ and $-|-67|$

(e) $-9$ and $9$

(f) $|-0.07|$ and $|0.07|$

(g) $-|-8|$ and $8$

(h) $-|-100|$ and $-100$

(i) $|0|$ and $-|0|$

(j) $|2-3|$ and $|3-2|$

(k) $|4-7|$ and $|4+7|$

2. Find these sums and differences.

(a) $8-9$

(b) $-2+7$

(c) $-12-(-4)$

(d) $15+(-5)$

(e) $11-(-17)$

(f) $-0.5-0.25$

(g) $-\frac{1}{2}+\left(-\frac{1}{3}\right)$

3. Find these products and quotients.

(a) $(-5) \div (-5)$

(b) $56 \div (-8)$

(c) $7 \times (-3)$

(d) $(-4) \times (-12)$

(e) $-45 \div 15$

(f) $-\left(-\frac{3}{4}\right) \div \frac{6}{2}$

4. Simplify these expressions.

(a) $-(2-7) \times |-2+5|$

(b) $|-9+4|^2$

(c) $\dfrac{3|11-7|-4}{-|7-11|+2}$

(d) $-2\Big|4-|5-9|\Big|+6$

# 15

# CHAPTER 15
## RATIOS AND PROPORTIONS

### DOOMED AS DOOMED CAN BE

It was just like old times. Rebecca, Jake, Kyle, and I were going over ratios at the café tables outside the Finer Diner. We were laughing and cracking jokes about baking babkas and spy names. But it wasn't the same.

Why? you ask. Because now I had a *serious* crush on Kyle Thomas. Before, it was kind of like a fantasy crush. Being a Johnny London groupie was fun; stalking Kyle was funny. But everything had changed. Kyle and I shared some powerful moments in the humid spy cave, in my 2001 Saturn, and at the scenic overlook.

"Hey, can you guys pass the sugar?" asked Jake.

Kyle and I reached for the bowl at the same time. I got there first, so his hand closed over mine. We both shot our hands back to our sides.

"Okay, fine, I'll just get it myself," Jake said.

I mean, Kyle's twenty-one and I'm only seventeen! He's *way* too old and mature for me, even if he is bad at math. Every bone in my body was telling me it wouldn't work. The stats were against us.* I mean, how many May-July romances work out? On average, I bet one in ten. Maybe one in twenty. And that's all beside the point if you consider what would happen if my parents find out. They'd poke Kyle with needle darts and lock me in my room for life.

It was hopeless. As a couple, we were doomed.

But maybe I was blowing the whole age thing out of proportion. Maybe a quick review of ratios and proportions would help me get my head straight.

* More on statistics in Chapter 16.

# RATIOS

A **ratio** is a type of comparison of two things–a comparison through division. They come up most often in real-world contexts, but mathematically speaking they behave exactly like fractions. Let's see how ratios work.

## THE BASICS

In our after-school remedial-math class, there are $15$ students–$6$ girls and $9$ boys. You can say that the ratio of girls to boys in our math class is $6$ to $9$. In symbols, the ratio is written with a colon:

The ratio of girls to boys is $6 : 9$.

The expression $6 : 9$ is read as "six to nine."

Here's the big idea: you can (and should) treat the ratio $6 : 9$ as another way of expressing the fraction $\frac{6}{9}$. (In fact, sometimes you'll see a ratio written like a fraction: "The ratio of girls to boys is $^6/_9$" or even "The ratio of girls to boys is $\frac{6}{9}$.")

And since $\frac{6}{9}$ is the same thing as $\frac{2}{3}$, you can also say,

The ratio of girls to boys is $2 : 3$.

People also say things like "In Shanon's math class, there are $2$ girls for every $3$ boys."

## RATIOS IN LOWEST TERMS

The ratio $2 : 3$ is in lowest terms, but the ratio $6 : 9$ isn't. Being in lowest terms works the same way for ratios as it does for the corresponding fractions $\frac{2}{3}$ and $\frac{6}{9}$.

To reduce a ratio, think of it as a fraction. For example, the ratio $4 : 8$ reduces to $1 : 2$ because the fraction $\frac{4}{8}$ reduces to $\frac{1}{2}$ when both top and bottom are divided by $4$. In fact, you can even do the reducing in ratio form:

$$\overset{1}{\cancel{4}} : \overset{2}{\cancel{8}} = 1 : 2.$$

Another example: the ratio $12 : 4$ reduces to $3 : 1$. That's because $\frac{12}{4}$ in lowest terms becomes $3$, which is the same thing as $\frac{3}{1}$.

### To reduce or not to reduce?

Remember what I said about reducing fractions in Chapter 7? You can do it or not do it, depending on what you want or what your teacher wants, lowest terms is not necessarily better, and so on.

That's because fractions are abstract numbers; the exact form that you need them in depends on context. Because ratios often come up when people talk about the real world—and because people's brains are limited and are better at dealing with smaller numbers—it often makes sense to reduce ratios to lowest terms.

## RATIO PROBLEMS

Ratios are just a concept, a way to think about things. To get more familiar with them, let's do a couple of sample problems.

### First problem: What's the ratio?

Main Street in Springfield houses $20$ businesses. Four of them are corporate franchises—Starbucks, Wal-Mart, Barnes & Noble, and Bank of America. The rest are small mom-and-pop outfits. What's the ratio of the number of small businesses to the number of large businesses on Springfield's Main Street?

First, let's rewrite all the information we know in a more organized fashion. We know two things:

$$\text{small businesses} + \text{large businesses} = 20$$

and

$$\text{large businesses} = 4.$$

And we need to know the ratio

$$\text{small businesses} : \text{large businesses}.$$

To find the ratio, we first need to find the number of small businesses. Since there are $20$ businesses total, and $4$ of them are large, that must mean that there are $20 - 4 = 16$ small businesses. Therefore,

$$\text{small businesses} : \text{large businesses} = 16 : 4,$$

which reduces to $4 : 1$.

So there are $4$ small businesses for every large business on Main Street.

### Keeping things straight

A key part of working with ratios is maintaining the order of the two things being compared. Since we're asked for the ratio of small to large businesses, we have to make sure that the number of small businesses comes first when we make the calculation. Otherwise you'll get a very different answer. For example, in the Main Street problem, the ratio of small to large businesses is $4 : 1$, but the ratio of large to small businesses is $4 : 16$, or $1 : 4$. (These are reciprocal ratios–remember reciprocals from Chapter 9?)

### Second problem: What's the component?

Rebecca's sock drawer is very messy–she just keeps her yellow socks and blue socks all jumbled up in a drawer. After doing laundry recently, her mother reported that the ratio of yellow socks to blue socks is $2 : 3$. Rebecca then rummaged in her sock drawer and counted $18$ blue socks. How many yellow socks does Rebecca have?

Again, let's rewrite what we know:

$$\text{yellow socks} : \text{blue socks} = 2 : 3$$

and

$$\text{blue socks} = 18.$$

One easy way to find the number of yellow socks is to use a technique called **cross-multiplying**. We already talked about it a bit in Chapter 7, and I'll explain why it's the same thing in a minute. First, rewrite the ratio as a fraction:

$$\frac{\text{yellow socks}}{\text{blue socks}} = \frac{2}{3}.$$

Put in any numbers that you know. Since we know that Rebecca has $18$ blue socks, we can rewrite:

$$\frac{\text{yellow socks}}{18} = \frac{2}{3}.$$

Next, cross-multiply: multiply the numerator of the first fraction by the denominator of the second and vice versa:

$$\frac{\text{yellow socks}}{18} \times \frac{2}{3}$$

$$(\text{yellow socks}) \times 3 = 2 \times 18$$

or

$$(\text{yellow socks}) \times 3 = 36.$$

Since $3$ times the number of yellow socks is $36$, there must be $36 \div 3 = 12$ yellow socks.

Let's check: the ratio of yellow socks to blue socks becomes $12 : 18$, which does indeed reduce to $2 : 3$ (divide both parts of the ratio by $6$).

### Why cross-multiplication works

In Chapter 7, we talked about using cross-multiplication to check whether two fractions were equivalent. If the two cross-products are the same, the fractions are two names for the same quantity.

The same thing is happening here—except in reverse. We have two fractions, $\frac{\text{yellow socks}}{18}$ and $\frac{2}{3}$, and we *know* that they're equal. Equal fractions give equal cross-products when cross-multiplied. So $\frac{\text{yellow socks}}{18} = \frac{2}{3}$ means that $(\text{yellow socks}) \times 3 = 2 \times 18$.

### Third problem: What are the *two* components?

This one is quite tricky, so I won't mind at all if you skip it.

My other friend Beth really likes to read and keeps borrowing my books. She's got $60$ books in her room, all of them either mine or hers. For every book that's mine, she has $3$ books that are hers. How many of the books in her room are mine?

Let's see what we can make of it. First, rewrite the information:

$$\text{total books} = 60$$

and

$$\text{my books} : \text{Beth's books} = 1 : 3.$$

Did you catch how "for every book that's mine, she has $3$ books that are hers" translates to "the ratio of my books to her books is $1 : 3$"?

Wouldn't it be great if we knew the ratio of my books to total books? Let's see what we can do.

Since there is $1$ of my books for every $3$ of Beth's books, there is $1$ of my books for every $4$ total books. This $4$-book total is made up of $1$ of my books and $3$ of Beth's books. So the ratio of my books to total books is $1 : 4$.

Let's convert to a fraction:

$$\frac{\text{my books}}{\text{total books}} = \frac{1}{4}.$$

Since

$$\text{total books} = 60,$$

we can rewrite:

$$\frac{\text{my books}}{60} = \frac{1}{4}.$$

Cross-multiply to get

$$(\text{my books}) \times 4 = 1 \times 60.$$

So

$$(\text{my books}) \times 4 = 60$$

and

$$\text{my books} = \frac{60}{4} = 15.$$

So $15$ of the $60$ books in Beth's room are mine. That means that the other $60 - 15 = 45$ are hers. Let's just check the ratio:

$$\frac{\text{my books}}{\text{Beth's books}} = \frac{15}{45},$$

which reduces to $\frac{1}{3}$. Perfect!

# PROPORTIONALITY

Proportions are a lot like ratios—it's all about expressing a relationship between things through dividing. But because people use the words *proportion* and *proportional* in plain speech all the time, a couple of related meanings have evolved. I want to go over all of them and show how they're related. I'll start with the one that often appears in math textbooks right after the discussion of ratios.

## EQUAL RATIOS

Mathematically, a **proportion** is an equality between two ratios. For example, the ratio of $2$ to $10$ is the same as the ratio of $3$ to $15$. (Check this: $2 : 10$ is the same thing as $1 : 5$. And $3 : 15$ is also the same thing as $1 : 5$.)

We can say that $2$ **is to** $10$ **as** $3$ **is to** $15$. This is a proportion. The first number in each pair is one-fifth of the second number.

You can express this proportion like this:

$$2 : 10 = 3 : 15$$

or like this:

$$\frac{2}{10} = \frac{3}{15}.$$

Here's another example: $16$ is to $12$ as $20$ is to $15$. That's because

$$\frac{16}{12} = \frac{20}{15}.$$

You can check that these two fractions are equivalent by cross-multiplying: $16 \times 15 = 240$ and $20 \times 12 = 240$, so the equality is true.

## CONSTANT OF PROPORTIONALITY

A proportion expresses the fact that two pairs of things are related in the same way. In the first example, the second number in each pair, divided by $5$ (or multiplied by $\frac{1}{5}$), gives the first number. In the second example, the second number in each pair has to be multiplied by $\frac{4}{3}$ to get the first number. (Check this: $12 \times \frac{4}{3} = 16$ and $15 \times \frac{4}{3} = 20$.)

This is always the way it works with proportions. The second number multiplied by the **constant of proportionality** gives the first number.

Let's look at the examples again. In the first example, the constant of proportionality is $\frac{1}{5}$. In the second, it's $\frac{4}{3}$. But $\frac{1}{5}$ is exactly the ratio of $2$ to $10$, or of $3$ to $15$. And $\frac{4}{3}$ is exactly the ratio $16 : 12$, or $20 : 15$.

The constant of proportionality is the ratio of the first number to the second.

## SAME RELATIONSHIP (AGAIN)

This is, like, the third time I'm saying this, but it's really true. A proportion says that the relationship between pairs of things is the same. In math, this works by multiplication by a constant number. But the same relationship exists outside of math as well.

For example, you can say that *man* is to *men* as *woman* is to *women*. The second thing in both pairs is the plural of the first.

Another language example: *go* is to *went* as *swim* is to *swam*. The second verb in both pairs is the past tense of the first.

## BEING PROPORTIONAL

That's all fine and good, you might say, but I don't see what equal ratios have to do with *two things* being proportional. You hear all the time things like "Pressure is proportional to temperature" or "The amount of money I spend on avocados each week is proportional to the number of avocados I buy" or even "My mother's anger is proportional to the number of dirty dishes in the sink."

One quantity is **proportional** to another if they increase and decrease together. Usually, we phrase the proportionality statement so that the first thing depends on the second. More temperature means more pressure; less temperature means less pressure. More avocados bought means more money spent; fewer avocados means less money. More dirty dishes means more rage and frustration; fewer dishes means less.

Mathematically, one quantity being proportional to another means that the relationship between these two quantities is always the same: there is a number—the constant of proportionality—by which you multiply the second number to get the first.

If I buy 2 avocados, I spend \$3. If I buy 4 avocados, I spend \$6. This makes a proportion:

$$\frac{\$3}{2 \text{ avocados}} = \frac{\$6}{4 \text{ avocados}}.$$

(**Check:** $3 \times 4 = 6 \times 2$. This really is a proportion.)

The constant of proportionality is $\frac{\$3}{2 \text{ avocados}}$, or $\frac{\$1.50}{1 \text{ avocado}}$. That's just the price of an avocado: \$1.50 per avocado.

## PROPORTIONALITY PROBLEMS

Problems on proportions are very similar to problems on ratios, so we'll just do a couple.

### First problem: Give the fourth number

18 is to 12 as 12 is to what number?

This is a very common textbook problem. Set up the proportion, paying careful attention to order. Don't be thrown off because the number 12 appears twice.

$$\frac{18}{12} = \frac{12}{\text{what number}}$$

Many people use a question mark (?) or a box (□) or a letter (*x*) in place of the number in the denominator. But it doesn't matter what you call it. Just cross-multiply:

$$18 \times (\text{number}) = 12 \times 12.$$

Simplify the right-hand side:

$$18 \times (\text{number}) = 144.$$

And finally undo the multiplication by dividing by 18:

$$\text{number} = 144 \div 18 = 8.$$

So 18 is to 12 as 12 is to 8.

### Second problem: Two proportional quantities

The number of dog-eared and ratty copies of *The Great Gatsby* that Maplewood High School throws out every year is roughly proportional to the number of juniors at the school. Last year, the school enrolled 228 juniors and ended up throwing out 36 copies of *The Great Gatsby*. This year, the junior class numbers 361. How many copies of F. Scott Fitzgerald's timeless masterpiece can school officials expect to retire this June?

Because the number of books thrown out is proportional to the number of juniors, we can set up a relationship:

$$\frac{\text{books tossed last year}}{\text{juniors last year}} = \frac{\text{books tossed this year}}{\text{juniors this year}}.$$

We know some of these numbers: 240 juniors and 36 books last year; 361 juniors this year. Put them in:

$$\frac{36}{228} = \frac{\text{books tossed this year}}{361}.$$

At this point, you can cross-multiply. However, I think I'd like to reduce the fraction on the left first—that will make the numbers smaller. Both 36 and 228 are divisible by 12, so

$$\frac{\overset{3}{\cancel{36}}}{\underset{19}{\cancel{228}}} = \frac{\text{books tossed this year}}{361}.$$

And now I cross-multiply:

$$3 \times 361 = (\text{books tossed this year}) \times 19.$$

And finally,

$$(\text{books tossed this year}) = \frac{3 \times 361}{19} = 57.$$

So Maplewood High's teachers can expect to chuck 57 copies of *The Great Gatsby*.

# YOUR TURN

Solutions start on page 308.

1. Find these ratios in lowest terms.

   (a) Kyle has the best music collection. (I know this because I borrowed his iPod last time we were in math class.) He says he has bought about $550$ CDs and downloaded another $220$. What's the ratio of CDs he has bought to CDs he has downloaded?

   (b) Becca's friend Emma Chestnut loves to read. Last year, she read $56$ nonfiction books and $72$ fiction books. What's the ratio of fiction books to nonfiction books that Emma read last year?

   (c) My mom is hoping that I'll start studying for the SATs soon, so she bought me a set of $600$ vocabulary flashcards. (My math score is great, of course, but my critical reading needs some serious work. And I shudder when I think of that new essay thing.) After weeks of being nagged, I finally went through the cards and found that I don't know a whopping $375$ of the words. What is the ratio of SAT words I know to SAT words I don't know on these flashcards?

2. Complete each proportion.

   (a) $2$ is to $5$ as $10$ is to ____
   (b) $8$ is to $28$ as ____ is to $84$
   (c) $9$ is to ____ as $75$ is to $175$
   (d) $6$ is to $4$ as $1.5$ is to ____

3. Solve these problems.

   (a) Becca was playing around with measuring tape and found that the ratio of the length of her arm to the length of her leg is $5 : 6$. If her arms are $25$ inches long, how long are her legs?

(b) Even though we live in Chicago, my mom is a huge Red Sox fan. (That's Boston's baseball team, in case you're an immigrant or something.) On average, they win $5$ games for every $8$ games they play. If they've played $100$ games so far, about how many games have they lost? (Oh, and if you're baseball-challenged: there are no ties in baseball.)

(c) Last night in the drippy cave, Dad's beautiful blond secretary was making cheesecake. The recipe says that you have to use $6$ packets of cream cheese and $2\frac{1}{4}$ cups of sugar, but she managed to scrounge together only $1\frac{1}{2}$ cups of sugar. How many packets of cream cheese should she use if she wants to keep the ingredients proportional?

(d) My uncle, who lives on a farm in Nebraska, says that the number of carrots you need to feed a bunch of bunnies is proportional to the number of bunnies you're feeding. (Duh.) If you need $36$ bushels of carrots to feed $15$ bunnies, how many bushels of carrots do you need to feed $20$ bunnies? How many bunnies can you feed with $60$ bushels of carrots?

4. Looking for his wallet, Jake turned out his pockets on the Finer Diner tables. He had $8$ pieces of pomegranate taffy, $16$ pieces of dill-flavored taffy, and $20$ pieces of caramel-popcorn taffy.

   We can say that the ratio of pomegranate taffy to dill taffy to caramel-popcorn taffy was $8$ to $16$ to $20$, or $8 : 16 : 20$.

   (a) Reduce this ratio to lowest terms.
   (b) Three hours later, Jake has $6$ pieces of pomegranate taffy left. If he has kept the ratio of pomegranate taffy to dill taffy to caramel-popcorn taffy the same, how many pieces of dill-flavored taffy does he have now? What about caramel-popcorn taffy?

5. My half-Mexican, half-Icelandic friend Emily P. has $30$ cousins. The ratio of the number of her Mexican cousins to the number of her Icelandic cousins is $7 : 3$. How many Mexican cousins and how many Icelandic cousins does my friend Emily P. have?

# 16

# CHAPTER 16
# STATISTICS AND GRAPHS



## KISSY-KISS

After our math session at the Finer Diner, I went home. During dinner, I poked at the eggplant mush and thought about telling my parents about me and Kyle. But I was just too exhausted to brace myself for the argument that would be sure to follow.

The next day was muggy and rainy. Classes passed in a blur, one after another. After school, I was supposed to go to the after-school remedial-math class and learn about statistics, but I was so depressed about how Kyle and I would never have the happiness that was rightfully ours that I was going to blow it off. Then someone tapped me on the shoulder at my locker.

"Kyle!" I exclaimed. "I didn't know you knew where my locker was!"

"I didn't, but I found out," said Kyle. "Can we talk?"

"Sure. I was just on my way to class," I lied.

"No, I mean, talk about us," said Kyle.

"Oh, um, sure, I guess," I said. "But I want to tell you up front—"

"Shh!" he said, and put his finger on my lips. My lips! He led me to an unused stairwell in our mammoth school building.

"Here's the deal," said Kyle. "You've found out a lot about me all on your own. You know I'm a spy called Agent Orange. You know that Johnny London is just a cover for me to become famous and then rule the world. But there's something you still don't know."

"What?" I asked, trying not to get lost in his baby-blue eyes. He has the longest, laziest, loveliest lashes!

"I'm not twenty-one," he said.

"Oh, super. So how old are you? Thirty-one? Fifty-one?"

"I'm seventeen. Just like you."

"But what about the beer at the Finer Diner?!"

"It was nonalcoholic," he said. "The waitress is a spy, too. She's in on the whole thing."

"So, you mean . . . ?"

"Shanon, I really like you. I was hoping we could be more than friends."

Then he kissed me! It was like the most intense I'd ever felt, squared.

"You are such an above-average kisser," I said to him.

"If I drew a line graph of how I feel, this moment would go off the charts, mate," he said to me.

Isn't it *so* cute how we always talk using math lingo? We make the best couple ever!

So we went to class.

* * * * * * * *

**Statistics** is the branch of math that deals with interpreting numerical data—quiz scores in your math class, for example. The more data you know, the more information you can squeeze out of it. One way of organizing lots of numerical information is visually, in a **graph**.

# DEFINITIONS

A list of numbers is a cumbersome thing. People look for ways to typify a list—to replace it with a number that preserves something of the spirit of the original list. There are a few common ways to come up with a number that keeps this spirit alive, and each number tells you something different.

Let's say we start by working with a list of seven numbers:

$$2, 2, 1, 8, 11, 4, 7.$$

## AVERAGE, OR MEAN

A mathematical **average** is obtained by adding all the numbers in the list and dividing by the number of numbers.

So the average of our list of seven numbers is

$$\frac{2 + 2 + 1 + 8 + 11 + 4 + 7}{7} = 5.$$

Another expression for *average* is **mean**. (The name is actually short for **arithmetic mean**.) The mean of our seven numbers is $5$.

If we replace every number in the list with $5$, the sum stays the same: $2 + 2 + 1 + 8 + 11 + 4 + 7$ is the same as $5 + 5 + 5 + 5 + 5 + 5 + 5$. This is one feature of averages.

## MEDIAN

The **median** is the middle-most number in a list. To find it, arrange the numbers in the list in order. The median is the one in the middle.

Our list, arranged in order, becomes

$$1, 2, 2, 4, 7, 8, 11.$$

Its median is $4$. That's the fourth number in our list of seven.

In any list, there are as many numbers less than the median as there are greater than the median. In this list, there are three numbers less than the median ($1$, $2$, $2$) and three numbers greater than the median ($7, 8, 11$).

If the list has an even number of entries, the median is the average of the middle two. So the median of $3$, $5$, $7$, and $8$ is $\frac{5+7}{2}$, which is $6$. We do this to preserve the property of medians that we just talked about–equal number of entries less than and greater than the median.

## MODE

The **mode** is the most frequently occurring number in a list.

The mode of our list is $2$.

In most cases, the mode is not a very useful way of conveying information about a list. Personally, I've always suspected that we only study it because the phrase "mean, median, and mode" sounds nice all together. Let's not waste any more time talking about it.

# INTERPRETATIONS

## WHAT DO THESE NUMBERS TELL US?

The mean and median together can tell you a bit about the list, but interpreting statistical quantities well takes practice.

For example, in our list, the median is $4$, and the mean is $5$. The median is a just little smaller than the mean. The fact that the mean and the median are pretty close together means that the set is reasonably well balanced–for every fairly large number (say, $11$) there is a fairly small number (say, $1$). The fact that the median is smaller than the mean signifies that the smaller numbers ($1, 2, 2$) are overall a little closer to the center than the larger numbers ($7, 8, 11$).

### ANOTHER EXAMPLE

Now, take a look at a different list:

$$1, 2, 2, 4, 7, 8, 3000.$$

The median is still $4$. But the mean is

$$\frac{1+2+2+4+7+8+3000}{7} = 432.$$

Because the median and the mean are so far apart, we know that this list is not balanced. The largeness of the large numbers isn't counterweighed by the smallness of the small numbers: $3000$ is a completely different beast from the rest of the group.

### IS THAT ALL?

Take a look at yet another list:

$$-2988, 2, 2, 4, 7, 8, 3000.$$

The median is $4$ again. The mean is

$$\frac{-2988+2+2+4+7+8+3000}{7} = 5.$$

These are the same as the values for mean and median that we got in the first list we looked at. This list is also relatively well balanced—the bigness of $3000$ is pacified by the smallness of $-2988$. But the list is still very different: in the first list, all the numbers were relatively close together. In this list, there is a very small number, a very large number, and five numbers sort of in the middle.

Evidently, the mean and the median aren't enough to characterize a list completely. We also need some kind of way to convey the range and spread of the numbers involved. Statistics can do all that, but some of the computations can get pretty involved.

# GRAPHS

Presenting information in picture form serves much the same goal as finding good statistical quantities: it gives you some idea of what's going on at a glance. A well-made graph can give you the flavor of a list very quickly—you know what they say about a picture being worth a thousand words.

Statistical pictures are called **graphs**, and there are a few common types.

## PICTOGRAPHS

A **pictograph** is one of the funnest and simplest types of graph. The best way to get to know it is to look at one. Johnny London had three weekend concerts, and Kyle made a pictograph of how many tickets were sold each night.

**Johnny London's ticket sales**

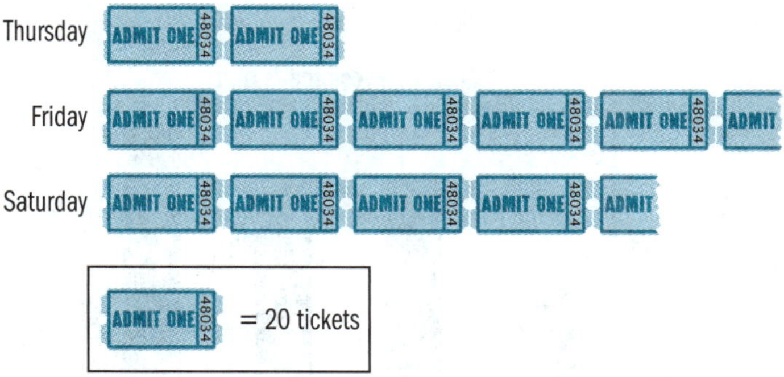


The text under the main graph tells us that each  stands for $20$ tickets. Two   next to "Thursday" means that Johnny London sold $2 \times 20 = 40$ tickets Thursday night. Half (  ) means $10$ tickets, so the five and a half  on Friday translate to $5 \times 20 + 10 = 110$ tickets. Four and half tickets on Saturday night means $90$ tickets.

The pictograph is exactly equivalent to three pieces of information: $40$ tickets on Thursday, $110$ on Friday, and $90$ on Saturday. But because the data is arranged visually, you can glean more information at a glance. Friday was the most popular night. Thursday was the least popular. Saturday is in between—it's the median. Moreover, you can tell something about the distribution: the difference between Thursday's and Saturday's sales is greater than the difference between Friday's and Saturday's.

All pictographs work the same way. A little picture (a **pictogram**) represents some kind of quantity, as explained somewhere nearby. This explanatory bit is called the **legend**. The number of pictures corresponds to the size of each entry.

## BAR GRAPHS

A **bar graph** is very similar to a pictograph—except that the information is represented with no-nonsense bars rather than cute pictures.

Today in class, Kyle, Jake, Rebecca, and I all got back our results on yesterday's math quiz. Take a look at the results.

The height of each person's bar corresponds to his or her quiz score. If the scores weren't labeled, you'd be able to estimate how well each person did by comparing the bar to the scale on the left. For example, Rebecca's bar is just over the $60\%$ tick. Jake's is just under the $80\%$ tick. Mine is close to the very top. And Kyle's . . . well, Kyle's is close to the halfway point between $40\%$ and $60\%$, so a bit worse than $50\%$. What can I say? It's not his math scores that make me swoon.

The bar graph conveys exactly four pieces of information: I scored $94\%$, Rebecca scored $62\%$, Jake scored $78\%$, and Kyle—bless his math-dummy, guitar-rocking soul—scored $47\%$. But you can also tell a number of things at a glance: Kyle got the lowest score; I got the highest score; no one got (much) more than half the quiz wrong.

### Bar graphs vs. pictographs

Any bar graph can be converted to a pictograph and vice versa. But pictographs work best when whatever is being represented is made up of separate, discrete units. So tickets or people work well—a picture of a ticket or a silhouette of a person can represent a specific number of tickets or people. But having a giant percentage sign representing $10\%$, or a picture of a clock representing $5$ hours, is a little silly.

## LINE GRAPHS

A **line graph** is a great way to track changes in data. Here's a line graph of Kyle's quiz scores over the last few days.

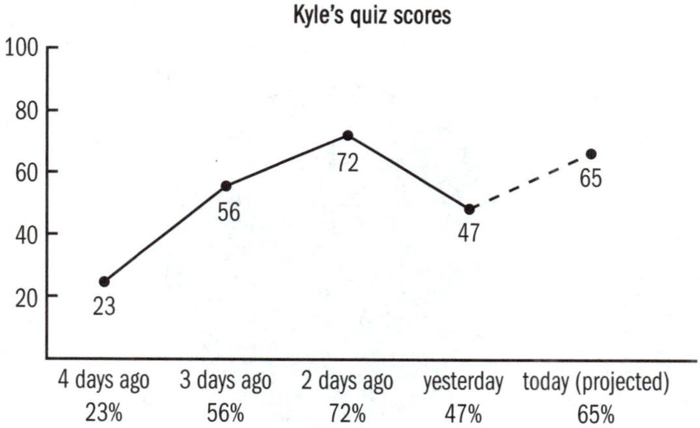


You can see the $47\%$ that Kyle got on yesterday's quiz, as well his scores on the three quizzes before that. The last entry isn't a *real* score—it's the score I *hope* he'll get on the quiz we have to take today.

The height of each dot corresponds to Kyle's score. Even if his scores weren't labeled, you could estimate them by looking at the scale on the left. The line that connects the dots makes it easy to see how his scores have changed. It tilts up when his score improves (as on the first three tests) and tilts down when his score falls (as it did on yesterday's quiz).

### Line graphs vs. bar graphs

Any line graph can be converted to a bar graph and vice versa. Line graphs make sense when the data points have some kind of connection between them—often, when they're representing something changing in time. Otherwise, drawing the line doesn't make much sense.

## PIE CHARTS

A **pie chart** (or a **pie graph**) is a great way to represent parts of a whole. In fact, we did similar things when we first talked about fractions, in Chapter 7. Here's a pie chart that visually shows how much of my time I spend thinking about different things.

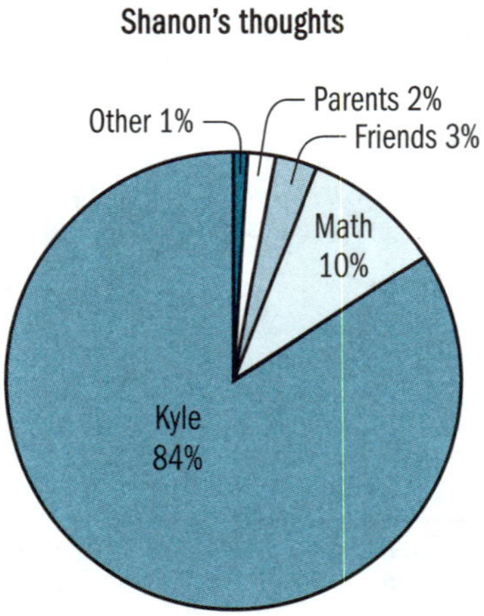


The bigger the slice that corresponds to something, the more time I spend thinking about it. You can see that I spend a *huge* portion of my time—more than three-quarters!—daydreaming about dreamy Kyle.

### Piecharts vs. other graphs

Pie charts are different from the other three types of graphs. A pie chart only works when everything you're representing adds up to some kind of whole pie. For example, the whole pie in the pie chart above is my brain-time.

# YOUR TURN

Solutions start on page 311.

1. Consider the list

$$67, 3, 8, 89, 10, 3, 100, 3, 54, 89, 9.$$

   (a) Find the median of these numbers.
   (b) Find the mean. Round to the nearest hundredth.
   (c) Does this set of numbers have a mode? If so, what is it?
   (d) If you add the number $101$ to the list, how does the mean change? Does the median change? Does the mode?

2. The average of six numbers is $44$. Five of the numbers are $46$, $50$, $34$, $20$, and $69$. What is the sixth number?

3. Johnny London made $\$1300$ from their series of three concerts. Kyle told me what they're planning to do with the money: $\$500$ to pay for renting the hall all three nights, $\$300$ to replace one of their crappy mikes, $\$80$ on new guitar picks, $\$50$ to pay for Jake's mom's parking ticket for Friday as they were unloading, $\$200$ for the after-party they had on Thursday, and the rest for Jake and Kyle and the other two band members to split. So we made a pie chart.

**Johnny London's proceeds**

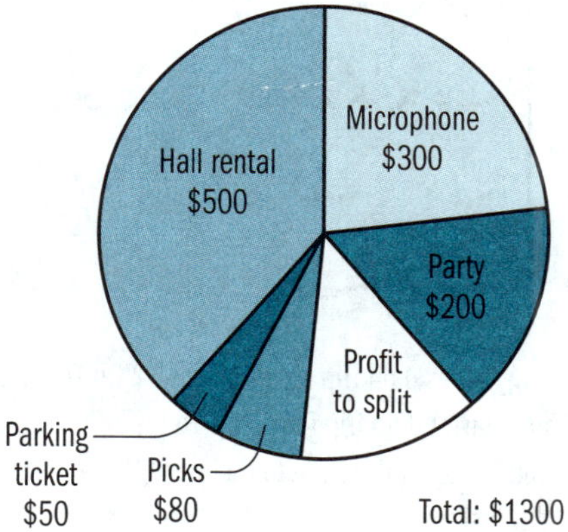

(a) What percentage of Johnny London's three-night earnings went to pay for the hall rental?
(b) Which single expense cost about a quarter of Johnny London's earnings?
(c) The chunk that the members of the band split is close to another expense. Which one? Is it more or less than that expense?
(d) The chunk that the band members split is close to two other expenses put together. Which ones? Are they more or less than the profit chunk?
(e) Use (c) and (d) to estimate the amount that the band members split.
(f) Calculate the amount that the band members split. How much money did Kyle make?

4. I don't like to talk about it, but I'm a huge klutz. I trip and fall down all the time. Rebecca helped me put together this pictograph of how many times I tripped last week.

**Shanon's trips**

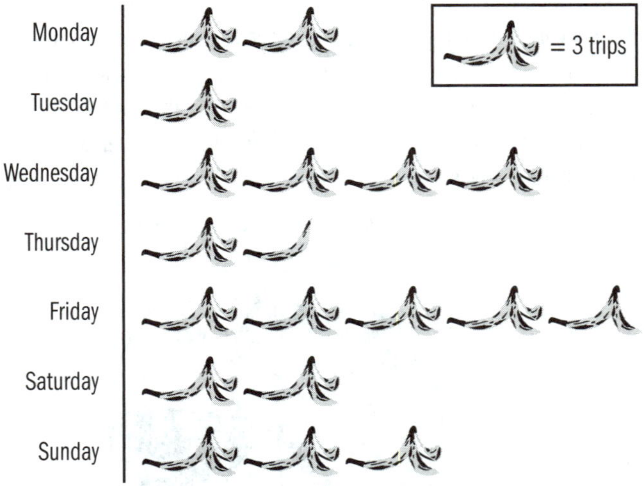


(a) How many times did I trip on Wednesday? On Thursday?
(b) On which day(s) did I trip fewer than 6 times?
(c) From the pictograph, estimate the average number of times I tripped each day last week.
(d) Compute this average.

5. My parents keep track of how many evil deeds each spy they're tracking commits. Here's the tally for this past month.

| **Spy Name** | **Number of evil deeds** |
|---|---|
| A. Gently | 23 |
| Agent Lee | 11 |
| Agent Orange (Kyle) | 16 |
| Agent Taff (Jake) | 7 |
| Black Mantis (Rebecca) | 41 |
| Boulder-dash | 8 |
| Ninjaman | 44 |
| Spy Child | 2 |

(a) Arrange this information in a bar graph. Use the bar graph to estimate the average number of evil deeds a spy tracked by my parents committed over the past month.

(b) What's the total number of evil deeds my parents tracked over the past month?

(c) Find the average number of evil deeds each spy committed. Were you close?

# REVIEW



## SECOND IMPRESSIONS

During our last remedial-math class, Kyle and I sat next to each other in the back row and passed notes back and forth. The coolest thing about falling for a singer-slash-songwriter-slash-guitar-player is that he writes a mean love letter! Lots of practice with love songs, you know.

After class, Rebecca and Jake wanted us to come to the Finer Diner, but I said no.

"Kyle and I have to go to my house. I promised my parents."

"You mean I have to meet your parents?" Kyle whined.

"You've already met them," I reminded him. "Remember, they drugged and jailed you and Jake in inhumane conditions?"

"No, I don't," Kyle said.

"Oh yeah, they gave you Memorase to block that all out. Sorry, I forgot."

Kyle was really nervous during the drive home. "I don't think I made a great first impression. They thought that I was trying to take over the world!"

"Can you blame them? You *were* trying to take over the world."

He insisted on stopping to get them flowers. By the time we walked through the front door, my parents were waiting for us in the living room.

"So, Agent Orange, we meet again," said my mom.

"Hello, Mrs. Fletcher," Kyle said. "I brought you some yellow carnations."

"Dusted with knockout powder?" my dad accused.

"No, Mr. Fletcher," Kyle said. "But I don't blame you for thinking bad thoughts about me. Please accept my sincerest apologies for spying and trying to end the world as you know it."

"Well, as long as he's sorry, honey," said my mom to my dad.

"Sorry doesn't always cut it," said my dad. "What are your intentions toward my daughter, Orange?"

"Please, call me Kyle–Kyle Thomas. That's my real name now. And my intentions are honorable. Through our spy games and our math tutoring sessions, I've fallen in love with your daughter. I want to make her very happy."

I inwardly squealed. *Kyle Thomas is in love with me!*

"Our sources say that you're twenty-one, but my seventeen-year-old daughter tells me that you're seventeen, too. Can you give me proof of your age?"

Kyle pulled out a birth certificate from Manchester General Hospital.

"Sir, please run this through all your spy networks. You will find that it is genuine."

"Oh, honey, don't be so hard on the boy," my mom said. "Thank you for coming over, Kyle. And I do want to apologize for injecting you with the Memorase and threatening you with Mongolian belly-button torture."

"Thank you, Mrs. Fletcher," Kyle said. We all waited for my dad to say something. He was looking over his shoulder, out the window into the backyard. When he turned around, I saw that he was crying! I'd never seen my dad cry before. It was weird. He came and gave me a big hug.

"My baby's growing up," he slobbered.

"Dad!" I exclaimed. But I hugged him back. He released me and turned to Kyle.

"If you ever harm a hair on her head, memory-erasing drugs and belly-button tortures will be the least of your worries," he warned. "Okay, now you kids go have fun. Here's ten dollars. Go see a movie. It's on me."

I don't know why my dad thinks ten bucks will pay for two movie tickets, but I took the bill.

"Thanks, Mr. Fletcher," Kyle said. We walked out of the house.

"Instead of seeing a movie, can we use the money to buy sodas and study math at the Finer Diner?" Kyle asked. "I couldn't concentrate at all today, but I think he was talking about algebra or something."

"You bet," I said.

And we did.

* * * * *

This isn't really a chapter. I mean, there's no new math to learn. But I have one problem for you, and it's a doozy. But make sure you do it.

## REVIEW PROBLEM

Today, Kyle, Jake, and I plotted Rebecca's mood swings over an hour. Every five minutes from $12{:}05$ p.m. to $1$ p.m., we asked her to rate how she was feeling: dreadful, mediocre, fine, pleasant, or glorious. Here's what we came up with. If I recall correctly, Jake stepped on her foot a little before $12{:}30$, but she recovered graciously.

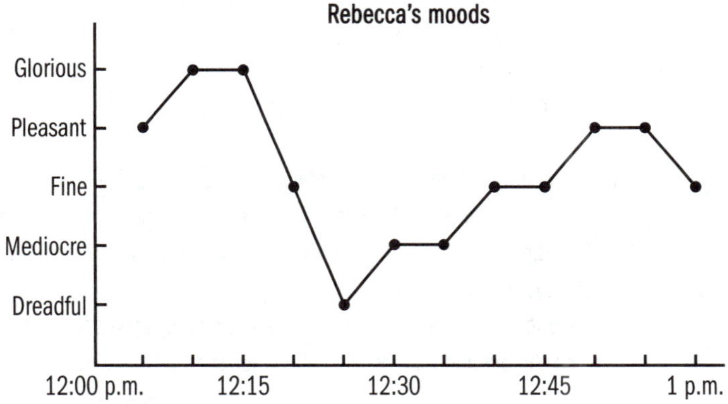


(a) How was Rebecca feeling at $12{:}15$? At $12{:}35$?

(b) List all the times when we recorded that Rebecca was feeling pleasant.

(c) During what period of time was Rebecca's mood dropping most quickly?

Next we converted Rebecca's moods to a five-point scale:

| | |
|---|---|
| Glorious | $2$ |
| Pleasant | $1$ |
| Fine | $0$ |
| Mediocre | $-1$ |
| Dreadful | $-2$ |

(d) Use this scale to find Rebecca's average mood over this hour.

REVIEW

Using our data, I made a table of how much time Rebecca spent in her various mood states.

| **Mood state** | **Time** |
|---|---|
| Glorious | 10 minutes |
| Pleasant | 20 minutes |
| Fine | 15 minutes |
| Mediocre | 10 minutes |
| Dreadful | 5 minutes |
| Total | 60 minutes |

(e) What percentage of her time did Rebecca spend feeling fine?

(f) What *fraction* of her time did Rebecca spend feeling glorious? Pleasant? Fine? Mediocre? Dreadful? Organize this information in a table. Check that all the fractions add up to 1.

(g) Use the fractions you found in (f) to draw a pie chart of Rebecca's mood states on the pie template below. Label each section.

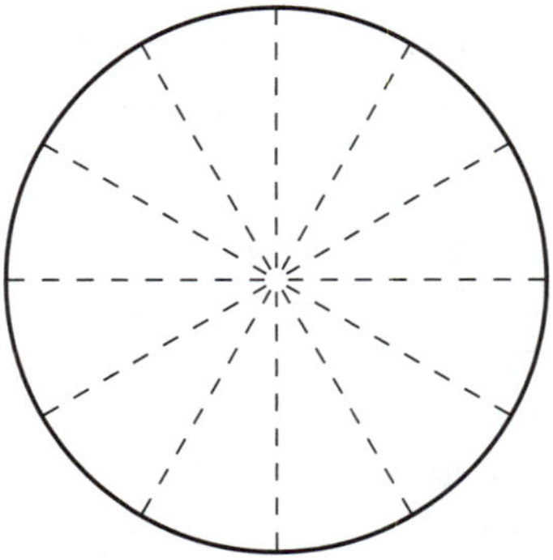



REVIEW

# EPILOGUE

## WHERE ARE THEY NOW?

So, that's my hundred-percent-true true story. Just in case you care, here's a quick update on what's happened in the last couple of months.

Rebecca and Jake are still going strong, although they recently had a nasty fight: Jake wants to open up a taffy factory in Maplewood, but Rebecca is afraid that he's going to balloon up from eating so much taffy and she'll have the fattest boyfriend in town. She's so vain!

Even if Jake ends up opening Taffywood (his name, not mine), it won't be for a while. Johnny London is releasing their debut album–*Spy Lies*–on Pfufferfish next week right after school lets out, so Jake's going to be busy touring this summer. So is Kyle, which makes me sad, but I'm going to tag along for a few shows when they come through Chicago, so that will totally rock. And Kyle has promised that he'll finish high school, so he'll be back in September.

My parents have given up the spy game. Dad is growing prize hydrangeas in the basement, and Mom is pursuing a second career in waste management. Oh, and Dad's beautiful blond secretary has moved into our garden shed. She's stopped bathing and asked us to call her Felt. Whatever.

Me? I'm still helping my friends with math and going to punk rock shows. Actually, I'm hoping to start tutoring for real this summer. If I can make enough money, my parents said I can fly out to see Johnny London in Manchester. That's in England! That's, like, five times more miles away than minutes I've spent kissing Kyle this last week!! I can't wait.

# ANSWER KEY

## CHAPTER 1

1. (a) 8
   (b) 6
   (c) 5

2. (a), (b), (c), and (f)
   Keep in mind that 0 is neither positive nor negative.

3. (a) 134,990
   (b) 135,000
   (c) 135,000
   (d) 130,000
   (e) 100,000
   Since 134,990 ends with 0, it's already rounded to the nearest ten. Also, did you notice? Rounding 134,990 to the nearest hundred and to the nearest thousand is the same thing.

4. One hundred and twenty-three billion, four hundred and fifty-six million, seven hundred and eighty-nine thousand, nine hundred and one.

5. 3,141,592,653

6. (a) True
   (b) False: 46 equals $56 - 10$, so it isn't greater than $56 - 10$.
   (c) True: 56 is greater than 46, so it is also *greater than* or equal to 46.
   (d) False: 46 equals 46, so it isn't less than 46.
   (e) True: 56 equals 56, so it is less than or *equal to* 56.

## CHAPTER 2

1. (c) and (e)
   The numbers in (a) and (b) are called *addends*. The expressions in (d) and (f) are nonsense.

2. $131 + 7245 = 7376$

Stack and add. No carrying necessary.

$$\begin{array}{r} 131 \\ +7245 \\ \hline 7376 \end{array}$$

3. $9273 + 3947 = 13{,}220$

Stack and add carefully, carrying where necessary.

$$\begin{array}{r} \scriptstyle 1\,1\,1\;\, \\ 9273 \\ +\ 3947 \\ \hline 13220 \end{array}$$

4. $927 - 82 = 845$

Stack and subtract, borrowing from the hundreds.

$$\begin{array}{r} \scriptstyle 8\;12\;\; \\ \cancel{9}\cancel{2}7 \\ -\ 82 \\ \hline 845 \end{array}$$

5. $301 - 127 = 174$

Stack and subtract, borrowing from the hundreds.

$$\begin{array}{r} \scriptstyle 2\;9\;11 \\ \cancel{3}\cancel{0}\cancel{1} \\ -127 \\ \hline 174 \end{array}$$

6. $16

Subtract the $34 cost of the book from the amount on the gift card before the purchase.

$$\begin{array}{r} \scriptstyle 4\;10 \\ \$\ \cancel{5}\cancel{0} \\ -\$\ 34 \\ \hline \$\ 16 \end{array}$$

7. 80 teachers

From the 85 tenured teachers last year, subtract the 8 who lost tenure: $85 - 8 = 77$. Then add the 3 teachers who got tenure: $77 + 3 = 80$.

8. 1103 students

Add up all the class sizes in a stack.

$$\begin{array}{r} \scriptstyle 2\,1\phantom{0} \\ 295 \\ 310 \\ 260 \\ 232 \\ +\ \ 6 \\ \hline 1103 \end{array}$$

9. 516

Since subtraction *undoes* addition, subtract 384 from 900 to find the missing number:

$$\begin{array}{r} \scriptstyle 8\ 9\ 10 \\ \cancel{9}\cancel{0}\cancel{0} \\ -384 \\ \hline 516 \end{array}$$

**Check:**

$$\begin{array}{r} \scriptstyle 1\,1\phantom{0} \\ 384 \\ +516 \\ \hline 900 \end{array}$$

## CHAPTER 3

1. (a) C
   (b) C
   (c) A
   (d) C
   (e) D
   (f) D

   Multiplying by 10 means adding on one 0 to the end of the number. Similarly, multiplying by 100 means adding on two 0s, and multiplying by 1000 means adding on three 0s.

2. (a) 70
(b) 100
(c) 110
(d) 85
(e) 1230

To multiply by 5 quickly, you can divide by 2 and then add on a 0 to the end.

3. 175

Stack and multiply. Stack multiplication is easier if the second number has fewer digits.

$$\begin{array}{r} \overset{3}{2}5 \\ \times\ 7 \\ \hline 175 \end{array}$$

4. 1331

Stack and multiply.

$$\begin{array}{r} 121 \\ \times\ 11 \\ \hline 121 \\ +121\ \\ \hline 1331 \end{array}$$

5. 361

Stack and multiply.

$$\begin{array}{r} 19 \\ \times\ 19 \\ \hline 171 \\ +\ 19\ \\ \hline 361 \end{array}$$

6. 172 minutes per week

Rebecca studies math 4 times a week, 43 minutes each time. That means she studies math a total of $43 \times 4$ minutes per week:

$$\begin{array}{r} {}^{1}\phantom{3} \\ 43 \\ \times\ 4 \\ \hline 172 \end{array}$$

7. 625,638

The 0 digit in 507 will result in an all-zero row. You can either keep that row in, as on the left, or take it out, keeping careful track of alignment, as on the right. Entirely up to you.

$$\begin{array}{r} 1234 \\ \times\ \ 507 \\ \hline 8638 \\ 0000\phantom{0} \\ +6170\phantom{00} \\ \hline 625638 \end{array} \qquad \begin{array}{r} 1234 \\ \times\ \ 507 \\ \hline 8638 \\ +6170\phantom{00} \\ \hline 625638 \end{array}$$

8. 1248 problems weekly!

If there are 32 kids and 13 problems per set per kid and 3 sets per week, then poor Mr. Honeyman has to grade $32 \times 13 \times 3$ problems per week. Multiply twice:

First, $32 \times 13 = 416$ problems per problem set.

$$\begin{array}{r} 32 \\ \times 13 \\ \hline 96 \\ +32\phantom{0} \\ \hline 416 \end{array}$$

And next, $416 \times 3 = 1248$ problems per week.

$$\begin{array}{r} {}^{1}\phantom{6} \\ 416 \\ \times 3 \\ \hline 1248 \end{array}$$

## CHAPTER 4

1. (a) 7, remainder 1
   (b) 8, remainder 3
   (c) 6, remainder 4
   (d) 8, remainder 3
   (e) 4, remainder 2
   (f) 7, remainder 4
   (g) 8, remainder 8

2. Rebecca is mistaken: $57 \div 11 = 5$, remainder 2.

   When dividing, the remainder always has to be less than the divisor. So just because Rebecca found a quasi-quotient and a quasi-remainder that work with the check

   divisor × quotient + remainder = original number

   doesn't mean she got the right answer.

   In fact, take a look below. I can find tons of quasi-quotients and quasi-remainders that work with the check:

$$11 \times 0 + 57 = 57$$
$$11 \times 1 + 46 = 57$$
$$11 \times 2 + 35 = 57$$
$$11 \times 3 + 24 = 57$$
$$11 \times 4 + 13 = 57$$
$$11 \times 5 + \phantom{0}2 = 57$$

   But the only quotient with a remainder less than 11 is 5. So that's the right one.

3. $467 \div 3 = 155$, remainder 2

$$\begin{array}{r} 155 \text{ R2} \\ 3\overline{)467} \\ -3\phantom{00} \\ \hline 16\phantom{0} \\ -15\phantom{0} \\ \hline 17 \\ -15 \\ \hline 2 \end{array}$$

4. $741 \div 7 = 105$, remainder 6

When you bring down the 4, the number you get (that's 4 again) is too small to divide by 7; you have to bring down the next digit (that's the 1) as well. You can do this either explicitly, as on the left, or in your head, as on the right. Either way, the fact that you need to bring down the next digit is recorded by the 0 digit in the answer.

$$\begin{array}{r} 105\ \text{R6} \\ 7\,\overline{)741} \\ -7\phantom{41} \\ \hline 04\phantom{1} \\ -00\phantom{1} \\ \hline 41 \\ -35 \\ \hline 6 \end{array} \qquad\qquad \begin{array}{r} 105\ \text{R6} \\ 7\,\overline{)741} \\ -7\phantom{41} \\ \hline 041 \\ -35 \\ \hline 6 \end{array}$$

5. (a) 56
   (b) 8
   (c) 7
   (d) 8
   (e) 7
   (f) 8
   (g) 4

6. 12 orders

Since I have \$50 and fries cost \$4 per order, I can afford roughly $50 \div 4$ orders of fries.

$$\begin{array}{r} 12\ \text{R2} \\ 4\,\overline{)50} \\ -4\phantom{0} \\ \hline 10 \\ -8 \\ \hline 2 \end{array}$$

7.
$$\begin{array}{r} 71\text{R}6 \\ 11\overline{)787} \\ \underline{-77} \\ 17 \\ \underline{-11} \\ 6 \end{array}$$

8. $10353 \div 17 = 609$

$$\begin{array}{r} 609 \\ 17\overline{)10353} \\ \underline{-102\phantom{0}} \\ 153 \\ \underline{-153} \\ 0 \end{array}$$

9. 87

If Jake divided properly, then the divisor times the quotient plus the remainder should give the original number:

| divisor × quotient | + | remainder | = original number |
|---|---|---|---|
| $7 \times 12$ | $+$ | $3$ | $= 87$ |

10. 28

We can do the same trick again:

$$\text{divisor} \times 6 + 0 = 168.$$

So $\text{divisor} \times 6 = 168$. That means that the divisor is $168 \div 6$.

$$\begin{array}{r} 28 \\ 6\overline{)168} \\ \underline{-12\phantom{0}} \\ 48 \\ \underline{-48} \\ 0 \end{array}$$

11. The short answer is, you can't divide by $0$.
More specifically, you can't say that $4 \div 0$ leaves a remainder of $4$ because *the remainder must be smaller than the divisor.* Since the divisor is $0$, no remainder will work—not even $0$!—because $0$ is the smallest whole number.

## CHAPTER 5

1. (a) $1$

Multiply & divide: $28 \div 4 - 2 \times 3 = 7 - 6$

Add & subtract: $7 - 6 = 1$

(b) $47$

Multiply & divide: $45 + 2 \times 7 - 12 = 45 + 14 - 12$

Add & subtract: $45 + 14 - 12 = 59 - 12$

$59 - 12 = 47$

(c) $16$

Parentheses multiply & divide:

$$34 \div (1 + 2 \times 2 \times 4) \times (56 \div 7) = 34 \div (1 + 4 \times 4) \times 8$$

$$34 \div (1 + 4 \times 4) \times 8 = 34 \div (1 + 16) \times 8$$

Parentheses add & subtract:

$$34 \div (1 + 16) \times 8 = 34 \div 17 \times 8$$

Multiply & divide: $34 \div 17 \times 8 = 2 \times 8$

$2 \times 8 = 16$

(d) $1$

Parentheses add & subtract:

$$13 + (15 - 14) \times 13 - 25 = 13 + 1 \times 13 - 25$$

Multiply & divide:

$$13 + 1 \times 13 - 25 = 13 + 13 - 25$$

Add & subtract: $13 + 13 - 25 = 26 - 25$

$26 - 25 = 1$

(e) 22

Nested parentheses:

$$(18 \times (12 \div 3) - (7 - 1)) \div (12 - (6 + 3)) = (18 \times 4 - 6) \div (12 - 9)$$

Parentheses multiply & divide:

$$(18 \times 4 - 6) \div (12 - 9) = (72 - 6) \div (12 - 9)$$

Parentheses add & subtract:

$$(72 - 6) \div (12 - 9) = 66 \div 3$$

Finish: $66 \div 3 = 22$

2. (a) 34
   (b) 0
   (c) No answer. You cannot divide by zero.
   (d) 17
   (e) 0
   (f) 0

3. (a) 2340

   By the distributive property (backwards),

   $234 \times 7 + 234 \times 3 = 234 \times (7 + 3) = 234 \times 10.$

   (b) 37

   By the commutative property of addition,

   $122 + 37 - 122 = 37 + 122 - 122.$

   Reparenthesize to $37 + (122 - 122) = 37 + 0.$

   (c) 2

   By the commutative property of multiplication,

   $145 \times 2 \div 145 = 2 \times 145 \div 145.$

   Reparenthesize to $2 \times (145 \div 145) = 2 \times 1.$

   (d) 7890

   By the distributive property (backwards),

   $12 \times 789 - 2 \times 789 = (12 - 2) \times 789 = 10 \times 789.$

   (e) 7373

   By the distributive property,

   $(100 + 1) \times 73 = 100 \times 73 + 1 \times 73 = 7300 + 73.$

(f) $340$

By the associative property of multiplication,

$(34 \times 5) \times 2 = 34 \times (5 \times 2) = 34 \times 10.$

(g) $1678$

By the associative property of addition,

$1 + (999 + 678) = (1 + 999) + 678 = 1000 + 678.$

4. (a) $9 \times 2 + 7 \times 2$
   (b) $9 \times 2 - 7 \times 2$
   (c) $2 \times 9 + 2 \times 7$
   (d) $2 \times 9 - 2 \times 7$

5. (a) $11 + 9 + 2$
   (b) $22 + 8 - 3$
   (c) $33 - 7 - 4$
   (d) $44 - 6 + 5$

6. (b), (e), and (f)
   (a) Nope. It's $5678$ that's being multiplied by $4567$, not $1234$.
   (b) Yes. In both, you're multiplying $5678$ and $4567$ and then adding $1234$.
   (c) Nope. Again, it's $5678$ that's being multiplied by $4567$ in the original expression, not $1234$.
   (d) Nope. In the original expression, only $5678$ gets multiplied by $4567$. In this one, you're also adding $1234 \times 4567$.
   (e) Yes. The expressions are identical—the parentheses here are only enforcing the standard order of operations.
   (f) Yes. Adding $0$ doesn't change anything.
   (g) No. Although flipping the order of multiplication and multiplying by $1$ doesn't change anything, multiplying $1234$ by $0$ sure does.

## CHAPTER 6

1. $15$ factors: $1, 2, 3, 4, 6, 8, 9, 12, 16, 18, 24, 36, 48, 72, 144$

2. Divisible by $2$: (a), (b), (e), and (f)

Numbers divisible by $2$ end in an even digit. Of (a) $2$, (b) $8$, (c) $7$, (d) $9$, (e) $0$, and (f) $2$, only $7$ and $9$ are *not* even.

Divisible by $3$: (a), (d), and (f)

Numbers divisible by $3$ have a digit sum divisible by $3$. Of (a) $27$, (b) $40$, (c) $26$, (d) $39$, (e) $34$, and (f) $21$, only $27$, $39$, and $21$ are divisible by $3$.

Divisible by $4$: (a) and (b)

A number is divisible by $4$ if its last two digits are divisible by $4$. Of (a) $72$, (b) $68$, (c) $27$, (d) $09$, (e) $10$, and (f) $02$, only $72$ and $68$ are divisible by $4$.

Divisible by $6$: (a) and (b)

A number is divisible by $6$ if it is divisible both by $2$ and by $3$. That's only (a) and (b).

Divisible by $12$: (a) only

By analogy with the rule for divisibility by $6$, a number is divisible by $12$ if it is divisible both by $4$ and by $3$. Only (a) matches that description.

3. (a) Prime. (Check divisibility by $2$, $3$, $5$, and $7$.)
   (b) Not prime. Divisible by $7$ (and by $13$).
   (c) Prime. (Check divisibility by $2$, $3$, $5$, $7$, and $11$.)
   (d) Prime. (Check divisibility by $2$, $3$, $5$, $7$, $11$, and $13$.)
   (e) Not prime. Divisible by $7$ (and by $11$ and $13$).

4. $144 = 2 \times 2 \times 2 \times 2 \times 3 \times 3$

There are many different factor trees. These are a few possibilities.

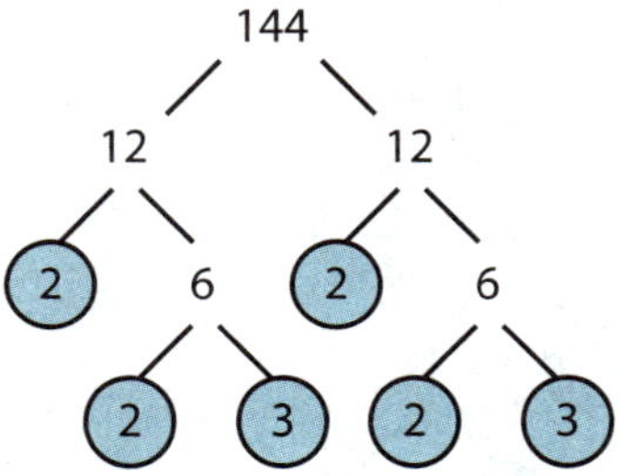


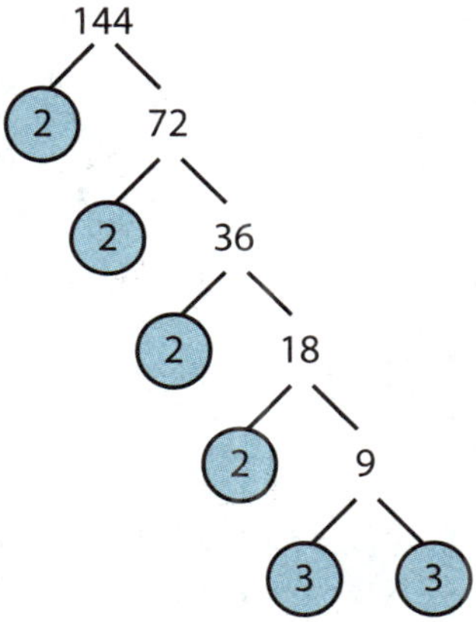


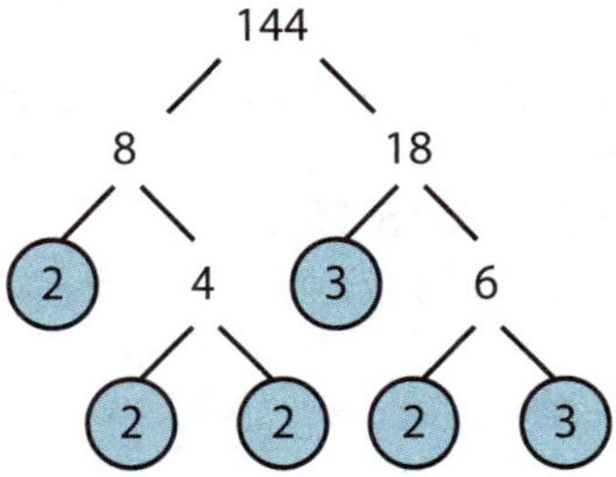


5. (a) $66 = 2 \times 3 \times 11$

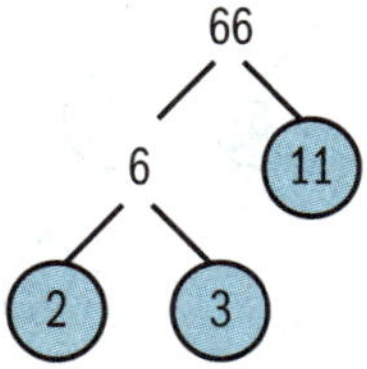


(b) $113$ is prime.

(c) $175 = 5 \times 5 \times 7$

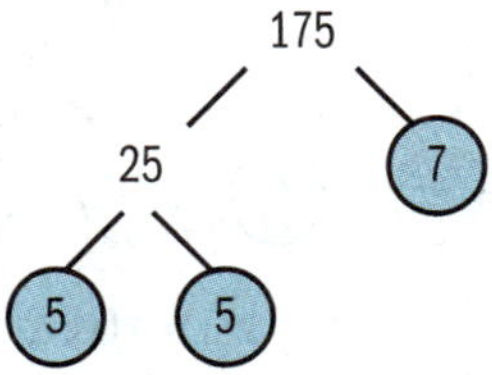

(d) $936 = 2 \times 2 \times 2 \times 3 \times 3 \times 13$

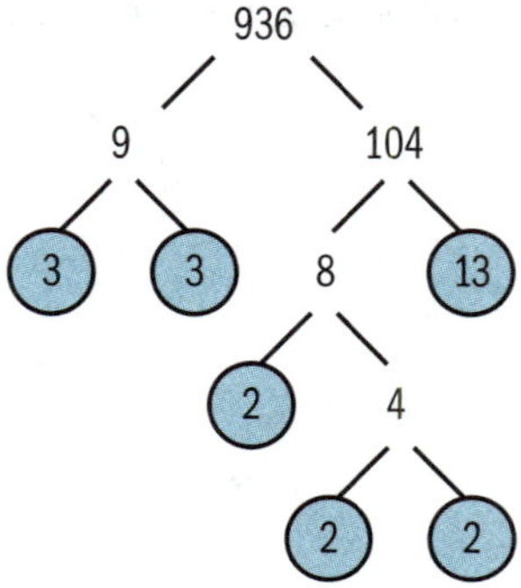


6. (a) Factors of $300$: $1, 2, 3, 4, 5, 6, 10, 12, 15, 20, 25, 30, 50, 60, 75, 100, 150, 300$
Factors of $225$: $1, 3, 5, 9, 15, 25, 45, 75, 225$
(b) Common factors: $1, 3, 5, 15, 25, 75$. The greatest common factor is $75$.
(c) $300 = 2 \times 2 \times 3 \times 5 \times 5$

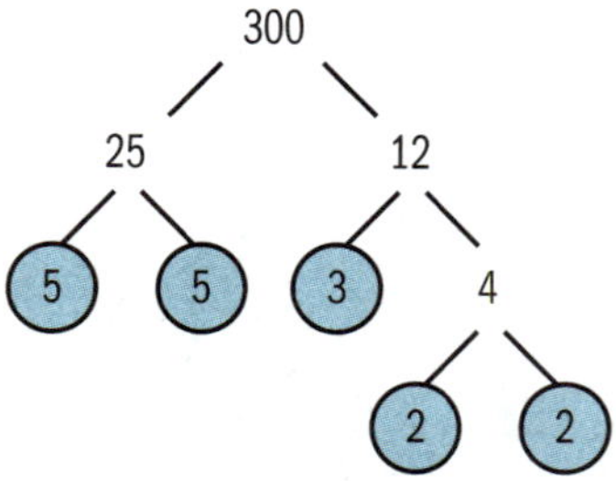


$225 = 3 \times 3 \times 5 \times 5$

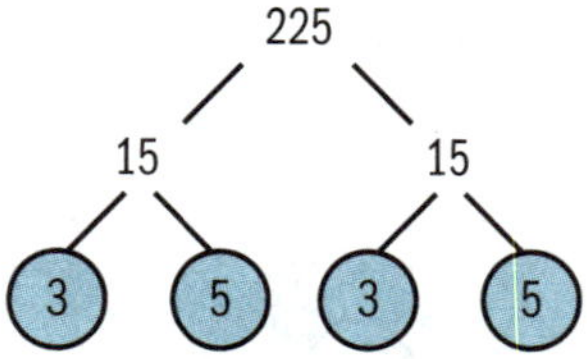


$300$ has two $2$s, one $3$, and two $5$s; $225$ has two $3$s and two $5$s. Therefore the GCF will have one $3$ and two $5$s. GCF $= 3 \times 5 \times 5$.

(d) $75 = 3 \times 5 \times 5$, so everything works out.

7. (a) $300$ has two $2$s, one $3$, and two $5$s; $225$ has two $3$s and two $5$s. Therefore the LCM will have two $2$s, two $3$s, and two $5$s.
LCM $= 2 \times 2 \times 3 \times 3 \times 5 \times 5 = 900$.

(b) LCM × GCF $= 75 \times 900 = 67{,}500$

(c) $300 \times 225 = 67{,}500$. The two products are equal, as expected.

8. (a) GCF is $1$. LCM is $85$.
(b) GCF is $12$. LCM is $60$.
(c) GCF is $7$. LCM is $5005$.

9. $84$

We know that the product of the two numbers is equal to the product of the GCF and the LCM, which is $21 \times 588 = 12{,}348$. Therefore the second number is $12{,}348 \div 147 = 84$. For fun, let's check. The prime factorization of $147$ is $3 \times 7 \times 7$. The prime factorization of $84$ is $2 \times 2 \times 3 \times 7$. Therefore their GCF has one $3$ and one $7$ (so $3 \times 7 = 21$) and their LCM has two $2$s, one $3$, and two $7$s (so $2 \times 2 \times 3 \times 7 \times 7 = 588$). Perfect.

10. (a) There are eight $2$s, four $3$s, and two $5$s in the prime factorization of $10!$. The prime factors are counted as follows.

$1$ contributes nothing
$2$ contributes one $2$
$3$ contributes one $3$
$4$ contributes two $2$s
$5$ contributes one $5$
$6$ contributes one $2$ and one $3$
$7$ contributes one $7$
$8$ contributes three $2$s
$9$ contributes two $3$s
$10$ contributes one $2$ and one $5$

(b) $10!$
$= 2 \times 2 \times 2 \times 2 \times 2 \times 2 \times 2 \times 2 \times 3 \times 3 \times 3 \times 3 \times 5 \times 5 \times 7$

This is often written as $2^8 \times 3^4 \times 5^2 \times 7$, in exponential notation. For more, see Chapter 12.

(c) $10!$ ends with $2$ zeros

End zeros come from factors of $10$, which are created from $2$s and $5$s. The prime factorization of $10!$ has eight $2$s but only two $5$s, so there will be only two factors of $10$ in $10!$. The actual number is $3{,}628{,}800$.

## CHAPTER 7

1. (a) Reduces to $\frac{1}{3}$

Both $3$ and $9$ are divisible by $3$, so $\frac{3}{9} = \frac{3 \div 3}{9 \div 3} = \frac{1}{3}$.

(b) In lowest terms

$2$ and $17$ don't have any common factors (both are prime).

(c) Reduces to $\frac{2}{9}$

The GCF of $16$ and $72$ is $8$, so $\frac{16}{72} = \frac{16 \div 8}{72 \div 8} = \frac{2}{9}$.

(d) Reduces to $\frac{3}{5}$

The GCF of $45$ and $75$ is $15$, so $\frac{45}{75} = \frac{45 \div 15}{75 \div 15} = \frac{3}{5}$.

(e) In lowest terms.

The prime factors of $30$ are $2$, $3$, and $5$, but $163$ isn't divisible by any of them. ($163$ isn't divisible by $2$ or $5$ because it ends in a $3$, and it isn't divisible by $3$ because its digits add up to $10$.) In fact, $163$ is prime.

(f) Reduces to $\frac{2}{19}$

$38 = 2 \times 19$, so we only have to check whether $361$ is divisible by $2$ or $19$. It turns out that $361 = 19 \times 19$. So $\frac{38}{361} = \frac{38 \div 19}{361 \div 19} = \frac{2}{19}$.

2. (b), (c), (d) and (f)

First, let's reduce $\frac{6}{14}$, just to simplify things: $\frac{6}{14} = \frac{6 \div 2}{14 \div 2} = \frac{3}{7}$.

That means that (a) $\frac{4}{7}$ doesn't work. To check the others, you can either reduce them to see if you also get $\frac{3}{7}$ or use the cross-multiplying tricky.

Reducing (b) $\frac{9}{21} = \frac{9 \div 3}{21 \div 3} = \frac{3}{7}$, (c) $\frac{33}{77} = \frac{33 \div 11}{77 \div 11} = \frac{3}{7}$, and (d) $\frac{12}{28} = \frac{12 \div 2}{28 \div 2} = \frac{6}{14}$ shows that all three do work. To check (e) and (f), let's use cross-multiplying. For (e), $222 \times 7 = 1554$ but $3 \times 504 = 1512$. Close but no cigar. But (f) works: $138 \times 7 = 966$ and $3 \times 322 = 966$.

3. (a) $1\frac{2}{4} = 1\frac{1}{2}$

Divide: $6 \div 4 = 1$, remainder $2$

(b) $2\frac{3}{9} = 2\frac{1}{3}$

Divide: $21 \div 9 = 2$, remainder $3$

(c) $8$

Divide: $144 \div 18 = 8$

(d) $3\frac{1}{7}$

Divide: $22 \div 7 = 3$, remainder $1$

(e) $3\frac{7}{11}$

Divide: $40 \div 11 = 3$, remainder $7$

4. (a) $\frac{5}{3}$

whole × denominator + numerator $= 1 \times 3 + 2 = 5$

(b) $\frac{8}{5}$

whole × denominator + numerator $= 1 \times 5 + 3 = 5$

(c) $\frac{11}{5}$

whole × denominator + numerator $= 2 \times 5 + 1 = 11$

(d) $\frac{13}{4}$

whole × denominator + numerator $= 3 \times 4 + 1 = 13$

(e) $\frac{131}{17}$

whole × denominator + numerator $= 7 \times 17 + 12 = 131$

5. (a) $\frac{1}{11}, \frac{2}{11}, \frac{6}{11}, \frac{7}{11}, \frac{9}{11}, \frac{11}{11}, \frac{15}{11}$

All of these fractions have the same denominator, so they increase as their numerators increase.

(b) $0, \frac{1}{11}, \frac{1}{7}, \frac{1}{4}, \frac{1}{3}, \frac{1}{2}, 1$

All of the fractions are greater than $0$ and less than $1$. The numerators are all equal, so the fractions increase as the denominators decrease.

(c) $\frac{1}{2}, \frac{4}{5}, \frac{8}{9}, \frac{10}{11}, \frac{11}{12}, 1$

All of the fractions are less than $1$. To put them in order, you might use the cross-multiplying trick. Or you might note that in every fraction, the numerator is one less than the denominator, which means that you can rewrite them as

$$\frac{11}{12} = 1 - \frac{1}{12},$$
$$\frac{4}{5} = 1 - \frac{1}{5},$$
$$\frac{1}{2} = 1 - \frac{1}{2},$$
$$\frac{10}{11} = 1 - \frac{1}{11}, \text{ and}$$
$$\frac{8}{9} = 1 - \frac{1}{9}.$$

$\frac{11}{12}$ is smaller than $\frac{10}{11}$, which is smaller than $\frac{8}{9}$, which is smaller than $\frac{4}{5}$, which is smaller than $\frac{1}{2}$. The leftovers when these fractions are subtracted from $1$ increase in reverse order.

(d) $\frac{2}{7}, \frac{1}{3}, \frac{2}{5}, \frac{3}{7}, \frac{1}{2}, \frac{4}{7}, \frac{3}{5}, \frac{2}{3}, \frac{5}{7}$

First, let's put the thirds and the half in order: $\frac{1}{3} < \frac{1}{2} < \frac{2}{3}$. Then insert the fifths, using cross-multiplying. Since $1 \times 5 < 2 \times 3$ but $2 \times 2 < 1 \times 5$, we know that $\frac{1}{3} < \frac{2}{5} < \frac{1}{2}$. Similarly, $\frac{1}{2} < \frac{3}{5} < \frac{2}{3}$. So $\frac{1}{3} < \frac{2}{5} < \frac{1}{2} < \frac{3}{5} < \frac{2}{3}$. Finally, let's insert the sevenths. Since $2 \times 3 < 1 \times 7$ we know that $\frac{2}{7} < \frac{1}{3}$. Similarly, $\frac{2}{3} < \frac{5}{7}$. Since $2 \times 7 < 3 \times 5$ but $3 \times 2 < 1 \times 7$ we know that $\frac{2}{5} < \frac{3}{7} < \frac{1}{2}$. Similarly, $\frac{1}{2} < \frac{4}{7} < \frac{3}{5}$. Therefore

$$\frac{2}{7} < \frac{1}{3} < \frac{2}{5} < \frac{3}{7} < \frac{1}{2} < \frac{4}{7} < \frac{3}{5} < \frac{2}{3} < \frac{5}{7}.$$

For good measure, here are all these fractions plotted on the number line:

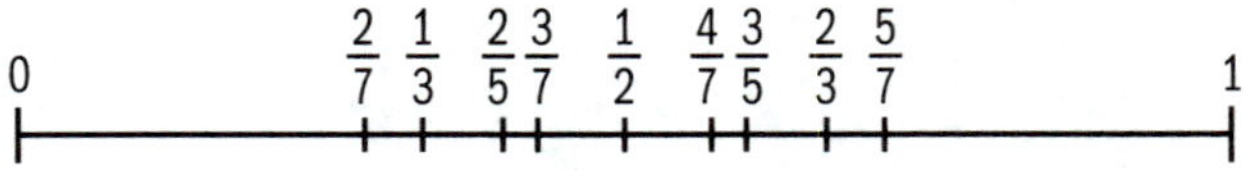

6. (a) less

Cross-multiply: $1 \times 8 < 5 \times 2$, so $\frac{1}{2} < \frac{5}{8}$.

(b) greater

Cross-multiply: $2 \times 8 > 5 \times 3$, so $\frac{2}{3} > \frac{5}{8}$.

(c) less

Cross-multiply: $3 \times 8 < 5 \times 5$, so $\frac{3}{5} < \frac{5}{8}$.

(d) less

Cross-multiply: $4 \times 8 < 5 \times 9$, so $\frac{4}{9} < \frac{5}{8}$.

(e) greater

Cross-multiply: $7 \times 8 > 5 \times 11$, so $\frac{7}{11} > \frac{5}{8}$.

(f) less

Cross-multiply: $8 \times 8 < 5 \times 13$, so $\frac{8}{13} < \frac{5}{8}$.

7. Ardent fans are a bigger group.

As we found in 5(d), so $\frac{2}{5} > \frac{1}{3}$.

8. Nothing.

I ate $\frac{1}{2}$ of the fries, and Rebecca ate $\frac{2}{4}$ of the fries. But $\frac{2}{4}$ is the same thing as $\frac{1}{2}$, so together we ate $\frac{1}{2} + \frac{1}{2}$, or one whole order.

## CHAPTER 8

1. (a) $\frac{4}{11}$

(b) $\frac{12}{18} = \frac{2}{3}$

(c) $\frac{20}{24} = \frac{5}{6}$

Compute: $\frac{21}{24} - \left(\frac{19}{24} - \frac{18}{24}\right) = \frac{21}{24} - \frac{1}{24} = \frac{20}{24}$

Reduce: $\frac{20}{24} = \frac{20 \div 4}{24 \div 4} = \frac{5}{6}$

2. (a) $\frac{8}{15}$

$$\frac{1}{3} + \frac{1}{5} = \frac{1 \times 5 + 1 \times 3}{3 \times 5} = \frac{8}{15}$$

(b) $\frac{70}{143}$

$$\frac{2}{11} + \frac{4}{13} = \frac{2 \times 13 + 4 \times 11}{11 \times 13} = \frac{70}{143}$$

(c) $\frac{5}{14}$

The LCM of $14$ and $7$ is $14$. $\frac{9}{14} - \frac{2}{7} = \frac{9 - 2 \times 2}{14} = \frac{5}{14}$

(d) $\frac{4}{5}$

The LCM of $8$ and $40$ is $40$.

Add: $\frac{5}{8} + \frac{7}{40} = \frac{5 \times 5 + 7}{40} = \frac{32}{40}$. Reduce: $\frac{32}{40} = \frac{32 \div 8}{40 \div 8} = \frac{4}{5}$

(e) $\frac{1}{84}$

The LCM of $21 = 3 \times 7$ and $28 = 2 \times 2 \times 7$ is $2 \times 2 \times 3 \times 7 = 84$.

$$\frac{1}{21} - \frac{1}{28} = \frac{1 \times 4 - 1 \times 3}{84} = \frac{1}{84}$$

(f) $\frac{11}{15}$

The LCM of $6 = 2 \times 3$ and $10 = 2 \times 5$ is $2 \times 3 \times 5 = 30$.

Subtract: $\frac{5}{6} - \frac{1}{10} = \frac{5 \times 5 - 1 \times 3}{30} = \frac{22}{30}$.

Reduce: $\frac{22}{30} = \frac{22 \div 2}{30 \div 2} = \frac{11}{15}$.

3. (a) $\frac{47}{60}$

The LCM of $3$, $4$, and $5$ is $60$. $\frac{1}{3} + \frac{1}{4} + \frac{1}{5} = \frac{1 \times 20 + 1 \times 15 + 1 \times 12}{3 \times 4 \times 5}$

$$= \frac{47}{60}$$

(b) $\frac{5}{6}$

The LCM of $2$ and $3$ is $6$. $1 - \left(\frac{1}{2} + \frac{1}{3}\right) = 1 - \frac{1 \times 3 + 1 \times 2}{6}$

$$= \frac{6}{6} - \frac{1}{6} = \frac{5}{6}$$

(c) $\frac{17}{1001}$

The LCM of $77 = 7 \times 11$, $143 = 11 \times 13$, and $91 = 7 \times 13$ is $7 \times 11 \times 13 = 1001$.

$$\frac{1}{77} - \frac{1}{143} + \frac{1}{91} = \frac{1 \times 13 - 1 \times 7 + 1 \times 11}{1001} = \frac{17}{1001}$$

(d) $\frac{355}{432}$

The LCM of $12 = 2 \times 2 \times 3$, $54 = 2 \times 3 \times 3 \times 3$ and $16 = 2 \times 2 \times 2 \times 2$ is $2 \times 2 \times 2 \times 2 \times 3 \times 3 \times 3 = 432$.

$$\frac{7}{12} + \frac{23}{54} - \frac{3}{16} = \frac{7 \times (2 \times 2 \times 3 \times 3) + 23 \times (2 \times 2 \times 2) - 3 \times (3 \times 3 \times 3)}{432}$$

$$= \frac{355}{432}$$

4. (a) $1\frac{1}{12}$

$$3\tfrac{1}{3} - 2\tfrac{1}{4} = 3 - 2 + \frac{1 \times 4 - 1 \times 3}{3 \times 4} = 1\tfrac{1}{12}$$

(b) $4\frac{11}{24}$

The LCM of 6 and 8 is 24, so $\frac{17}{6} + \frac{13}{8} = \frac{17 \times 4 + 13 \times 3}{24} = \frac{107}{24}$.

Since $96 = 24 \times 4$ and $107 - 96 = 11$, $\frac{107}{24} = 4\frac{11}{24}$.

(c) $\frac{17}{18}$

$4\frac{1}{9} - \frac{19}{6} = 4\frac{1}{9} - 3\frac{1}{6}$. Since $\frac{1}{9}$ is less than $\frac{1}{6}$, we convert $4\frac{1}{9}$ to $3\frac{10}{9}$.

Now, $4\frac{1}{9} - 3\frac{1}{6} = 3\frac{10}{9} - 3\frac{1}{6} = \frac{10}{9} - \frac{1}{6}$. Finally, the LCM of 9 and 6 is 36, so $\frac{10}{9} - \frac{1}{6} = \frac{10 \times 4 - 1 \times 6}{36} = \frac{34}{36} = \frac{17}{18}$.

5. (a) $\frac{49}{24}$

The LCM of 12 and 8 is 24, so $\frac{11}{12} + \frac{9}{8} = \frac{11 \times 2 + 9 \times 3}{24} = \frac{49}{24}$.

(b) $\frac{31}{6}$

The LCM of $2$ and $3$ is $6$, so $3\frac{1}{2}+1\frac{2}{3} = \frac{7}{2}+\frac{5}{3} = \frac{7\times 3+5\times 2}{6} = \frac{31}{6}$.

(c) $\frac{433}{132}$

$$4\frac{7}{22}-\left(2\frac{1}{4}-\frac{40}{33}\right) = \frac{95}{22}-\left(\frac{9}{4}-\frac{40}{33}\right)$$

The LCM of $22 = 2\times 11$, $4 = 2\times 2$, and $33 = 3\times 11$

is $2\times 2\times 3\times 11 = 132$,

so $\frac{95}{22}-\left(\frac{9}{4}-\frac{40}{33}\right) = \frac{95\times 6-(9\times 33-40\times 4)}{132} = \frac{433}{132}$. Since $433$ is not divisible by $2$, $3$, or $11$, the fraction $\frac{433}{132}$ is in lowest terms.

## CHAPTER 9

1. (a) $\frac{7}{4}$

   (b) $\frac{6}{2}=3$

   (c) No answer. 0 doesn't have a reciprocal.

   (d) 1

   (e) $\frac{5}{6}$

   Convert $1\frac{1}{5}$ to improper fraction $\frac{6}{5}$ before flipping.

   (f) $\frac{1}{9}$

2. (a) $\frac{3}{4}$

   (b) $\frac{10}{21}$

   (c) $\frac{4}{9}$

   (d) $\frac{1}{512}$

   (e) $\frac{15}{56}$

3. (a) $\frac{2}{15}$

$$\frac{\cancel{3}^{1}}{\cancel{9}_{3}} \times \frac{2}{5} = \frac{2}{15}$$

(b) $\frac{3}{7}$

$$\frac{9}{\cancel{49}_{7}} \times \frac{\cancel{7}^{1}}{3} = \frac{\cancel{9}^{3}}{7} \times \frac{1}{\cancel{3}_{1}} = \frac{3}{7}$$

(c) $\frac{1}{5}$

$$\frac{3}{\cancel{60}_{15}} \times \cancel{4}^{1} = \frac{\cancel{3}^{1}}{\cancel{15}_{5}} = \frac{1}{5}$$

(d) $\frac{35}{3}$

Convert $2\frac{1}{3}$ to an improper fraction before multiplying: $5 \times \frac{7}{3} = \frac{35}{3}$

(e) $\frac{1}{72}$

$$\frac{3}{28} \times \frac{7}{\cancel{30}_{2}} \times \frac{\cancel{15}^{1}}{27} = \frac{3}{\cancel{28}_{4}} \times \frac{\cancel{7}^{1}}{2} \times \frac{1}{27} = \frac{\cancel{3}^{1}}{4} \times \frac{1}{2} \times \frac{1}{\cancel{27}_{9}} = \frac{1}{72}$$

(f) $\frac{1}{10}$

The denominator of each fraction (except for the last one) cancels with the numerator of the next fraction:

$$\frac{1}{2} \times \frac{2}{3} \times \frac{3}{4} \times \frac{4}{5} \times \frac{5}{6} \times \frac{6}{7} \times \frac{7}{8} \times \frac{8}{9} \times \frac{9}{10}$$

$$= \frac{1}{\cancel{2}} \times \frac{\cancel{2}}{\cancel{3}} \times \frac{\cancel{3}}{\cancel{4}} \times \frac{\cancel{4}}{\cancel{5}} \times \frac{\cancel{5}}{\cancel{6}} \times \frac{\cancel{6}}{\cancel{7}} \times \frac{\cancel{7}}{\cancel{8}} \times \frac{\cancel{8}}{\cancel{9}} \times \frac{\cancel{9}}{10} = \frac{1}{10}$$

4. (a) $\frac{1}{7}$

$$\frac{3}{7} \div 3 = \frac{3}{7} \times \frac{1}{3} = \frac{\cancel{3}}{7} \times \frac{1}{\cancel{3}} = \frac{1}{7}$$

(b) $\frac{1}{10}$

$$\frac{1}{5} \div 2 = \frac{1}{5} \times \frac{1}{2} = \frac{1}{10}$$

(c) $\frac{1}{3}$

$$\frac{1}{6} \div \frac{1}{2} = \frac{1}{6} \times \frac{2}{1} = \frac{1}{\underset{3}{\cancel{6}}} \times \frac{\cancel{2}}{1} = \frac{1}{3}$$

(d) 21

$$7 \div \frac{1}{3} = 7 \times \frac{3}{1} = 21$$

(e) 2

$$\frac{3}{2} \div \frac{3}{4} = \frac{3}{2} \times \frac{4}{3} = \frac{\cancel{3}}{\cancel{2}} \times \frac{\overset{2}{\cancel{4}}}{\cancel{3}} = 2$$

(f) $\frac{3}{8}$

$$\frac{1}{4} \div \frac{2}{3} = \frac{1}{4} \times \frac{3}{2} = \frac{3}{8}$$

(g) $\frac{26}{15}$

$$\frac{12}{25} \div \frac{18}{65} = \frac{\overset{2}{\cancel{12}}}{\underset{5}{\cancel{25}}} \times \frac{\overset{13}{\cancel{65}}}{\underset{3}{\cancel{18}}} = \frac{26}{15}$$

(h) 6

Convert $3\frac{3}{4} \div \frac{5}{8}$ into an improper fraction before dividing:

$$\frac{15}{4} \div \frac{5}{8} = \frac{15}{4} \times \frac{8}{5} = \frac{\overset{3}{\cancel{15}}}{\cancel{4}} \times \frac{\overset{2}{\cancel{8}}}{\cancel{5}} = 6.$$

5. (a) $\frac{1}{15}$

$$\frac{3}{4} \div \frac{9}{2} \times \frac{2}{5} = \frac{\cancel{3}}{\cancel{4}} \times \frac{\cancel{2}}{\underset{3}{\cancel{9}}} \times \frac{\cancel{2}}{5} = \frac{1}{15}$$

(b) $\frac{1}{5}$

$$\left(\frac{1}{5}+\frac{1}{10}\right)\times\frac{2}{3} = \left(\frac{2}{10}+\frac{1}{10}\right)\times\frac{2}{3} = \frac{\cancel{3}}{\cancel{10}_{5}}\times\frac{\cancel{2}}{\cancel{3}} = \frac{1}{5}$$

(c) $\frac{2}{3}$

$$\frac{5}{9}\div\left(\frac{3}{2}-\frac{2}{3}\right) = \frac{5}{9}\div\left(\frac{3\times3-2\times2}{2\times3}\right) = \frac{5}{9}\div\frac{5}{6} = \frac{\cancel{5}}{\cancel{9}_{3}}\times\frac{\cancel{6}^{2}}{\cancel{5}} = \frac{2}{3}$$

(d) $\frac{5}{2}$

$$\left(1+\frac{1}{2}\right)\left(1+\frac{1}{3}\right)\left(1+\frac{1}{4}\right) = \left(\frac{2}{2}+\frac{1}{2}\right)\times\left(\frac{3}{3}+\frac{1}{3}\right)\times\left(\frac{4}{4}+\frac{1}{4}\right)$$
$$= \frac{\cancel{3}}{2}\times\frac{\cancel{4}}{\cancel{3}}\times\frac{5}{\cancel{4}} = \frac{5}{2}$$

(e) 2

This is what's called a nested fraction. Never fear. The key thing to keep in mind is that a fraction is a division in progress:

$\dfrac{\frac{5}{3}-1}{2-\frac{5}{3}} = \left(\frac{5}{3}-1\right)\div\left(2-\frac{5}{3}\right)$. Now you can do the math:

$$\left(\frac{5}{3}-1\right)\div\left(2-\frac{5}{3}\right) = \left(\frac{5}{3}-\frac{3}{3}\right)\div\left(\frac{6}{3}-\frac{5}{3}\right) = \frac{2}{3}\div\frac{1}{3} = \frac{2}{3}\times\frac{3}{1} = 2.$$

## CHAPTER 10

1. (a) 4
   (b) 2
   (c) 7
   (d) 1
   (e) 5
   (f) 1

2. (a) 1
   (b) 0.1
   (c) 1.0
   (d) 0.200

3. (a) 4.589

$$\begin{array}{r} 3.000 \\ +1.589 \\ \hline 4.589 \end{array}$$

(b) 6.01

$$\begin{array}{r} 2.19 \\ +3.82 \\ \hline 6.01 \end{array}$$

(c) 2.96

$$\begin{array}{r} 4.05 \\ -1.09 \\ \hline 2.96 \end{array}$$

(d) 14.425

$$\begin{array}{r} 10.300 \\ +\ \ 4.125 \\ \hline 14.425 \end{array}$$

(e) 6.55

$$\begin{array}{r} 7.00 \\ -0.45 \\ \hline 6.55 \end{array}$$

(f) 7.11

$$\begin{array}{r} 9.20 \\ -2.09 \\ \hline 7.11 \end{array}$$

4. (a) 6.78
(b) 60
(c) 0.03809
(d) 87
(e) 105

5. (a) 6

$$\begin{array}{r} 12 \\ \times .5 \\ \hline 6.0 \end{array}$$

(b) 4.25

$$\begin{array}{r} .25 \\ \times\ 17 \\ \hline 175 \\ +25\phantom{0} \\ \hline 4.25 \end{array}$$

(c) 20

It's easier to multiply by a number with fewer digits, so flip the order.

$$\begin{array}{r} .625 \\ \times\ \ 32 \\ \hline 1250 \\ +1875\phantom{0} \\ \hline 20.000 \end{array}$$

(d) 99.9999

$$\begin{array}{r} 142.857 \\ \times\ \ \ \ \ .7 \\ \hline 99.9999 \end{array}$$

(e) 0.204222

$$\begin{array}{r} .00202 \\ \times\ 101.1 \\ \hline 202 \\ 202\phantom{0} \\ +202\phantom{00} \\ \hline .204222 \end{array}$$

6. (a) 7.2

Divide:

$$\begin{array}{r} 7.2 \\ 5\,\overline{)\,36.0} \\ -35\phantom{.0} \\ \hline 1\ 0 \\ -1\ 0 \\ \hline 0 \end{array}$$

Alternatively, you can determine that $36 \div 5 = 7$, remainder 1, which means that $36 \div 5 = 7\frac{1}{5}$. Finally, $\frac{1}{5} = 0.2$, so $7\frac{1}{5} = 7.2$.

(b) 0.04

Divide:

$$\begin{array}{r} 0.04 \\ 25\overline{)1.00} \\ -1\ 00 \\ \hline 0 \end{array}$$

(c) 0.9

Divide:

$$\begin{array}{r} 0.9 \\ 8\overline{)7.2} \\ -7\ 2 \\ \hline 0 \end{array}$$

Alternatively, you know that $72 \div 8 = 9$. Since $7.2$ is just a little less than $8$, you expect $7.2 \div 8$ to be just a little less than $1$. This corresponds to $0.9$ (as opposed to $90$ or $9$ or $0.09$).

(d) 0.65

Divide:

$$\begin{array}{r} 0.65 \\ 14\overline{)9.10} \\ -8\ 4 \\ \hline 70 \\ -70 \\ \hline 0 \end{array}$$

Since $9.1$ is more than $1.4$ but less than $14$, the quotient will be between $0.1$ and $1$. So $0.65$ fits best (rather than $6.5$ or $0.65$).

(e) 8

Convert $5.6 \div 0.7$ to $56 \div 7$. Then divide $56 \div 7 = 8$. Alternatively, you can think about it this way: since $5.6$ is more than $0.7$ but less than $7$, the quotient will be between $1$ and $10$. So $8$ fits (rather than $80$ or $0.8$).

(f) $1.1$

Convert: $1.21 \div 1.1$ is the same thing as $12.1 \div 11$. Then divide:

$$\begin{array}{r} 1.1 \\ 11\overline{)12.1} \\ -11\phantom{.1} \\ \hline 1\,1 \\ -1\,1 \\ \hline 0 \end{array}$$

Alternatively, simply find that $121 \div 11 = 11$. Since $1.21$ is just a little more than $1.1$, the quotient should be just a little more than $1$. This corresponds to $1.1$ (as opposed to $110$ or $11$ or $0.11$).

(g) $250$

Convert: $30 \div 0.12$ is the same thing as $3000 \div 12$. Then divide. Be sure not to lose the final zero—after you get the first digit, you know that the answer is a three-digit number:

$$\begin{array}{r} 250 \\ 12\overline{)3000} \\ -24\phantom{00} \\ \hline 60\phantom{0} \\ -60\phantom{0} \\ \hline 0\phantom{0} \end{array}$$

**Check:** since $30$ is more than $12$ but less than $120$, the answer will be between $100$ and $1000$. So $250$ works best (rather than $2.5$ or $25$ or $2500$).

7. (a) $\frac{9}{10}$

(b) $\frac{7}{100}$

(c) $\frac{12}{5}$

There are two ways to do this. You can convert $0.4$ to $\frac{4}{10} = \frac{2}{5}$, add $2$ to make $2\frac{2}{5}$, and convert $2\frac{2}{5}$ to an improper fraction. Alternatively, you can also say that $2.4$ is the same thing as $\frac{24}{10}$ (after all, if you divide $24 \div 10$ you get exactly $2.4$), and reduce $\frac{24}{10}$ to $\frac{12}{5}$.

(d) $\frac{7}{8}$

$\frac{875}{1000}$ reduces to $\frac{7}{8}$ when you divide top and bottom by $25$.

(e) $\frac{4}{3}$

You know that $0.33333...$ is $\frac{1}{3}$. Therefore $1.33333...$ is $1\frac{1}{3}$, or $\frac{4}{3}$.

(f) $\frac{1}{16}$

$\frac{625}{10000}$ reduces to $\frac{1}{16}$ when you divide top and bottom by $625$.

8. (a) $0.37$

(b) $0.6$

Divide:

$$\begin{array}{r} 0.6 \\ 5\,\overline{)3.0} \\ -3\,0 \\ \hline 0 \end{array}$$

Alternatively, if you happen to know by heart that $\frac{1}{5} = 0.2$, you can deduce that $\frac{3}{5} = 0.6$.

(c) $0.35$

Divide:

$$\begin{array}{r} 0.35 \\ 20\overline{)7.00} \\ -6\,0 \\ \hline 1\,00 \\ -1\,00 \\ \hline 0 \end{array}$$

Alternatively, if you happen to know that $\frac{1}{20} = 0.05$, you can multiply by 7 to get $\frac{7}{20} = 0.35$.

(d) $0.416666...$

Divide:

$$\begin{array}{l} \phantom{12)}0.4166\cdots \\ 12\overline{)5.0000}\ldots \\ \phantom{1}-4\,8 \\ \hline \phantom{12)1}20 \\ \phantom{12)}-12 \\ \hline \phantom{12)11}80 \\ \phantom{12)1}-72 \\ \hline \phantom{12)111}80 \\ \phantom{12)11}-72 \\ \hline \phantom{12)1111}8\ \ldots \end{array}$$

9. (a) V
   (b) IV
   (c) I
   (d) IX
   (e) II
   (f) III

(g) VII
(h) VIII
(i) VI

The easiest way to do this problem is to put the decimals in order and match them up with the fractions, least to greatest.

## CHAPTER 11

1. (a) $0.3$
   (b) $0$
   (c) $0.04$
   (d) $1.1$
   (e) $0.005$
   (f) $0.0125$
   First, convert to a decimal percentage: $1\frac{1}{4}\% = 1.25\%$. Then convert to a decimal.

2. (a) $37\%$
   (b) $102\%$
   (c) $0.6\%$
   (d) $450\%$
   (e) $70\%$
   First, convert to a decimal: $\frac{7}{10} = 0.7$. Then convert to a percentage.
   (f) $18.75\%$
   First, convert $\frac{3}{16}$ to a decimal.

$$\begin{array}{r} 0.1875 \\ 16\,\overline{)3.0000} \\ -1\,6\phantom{000} \\ \hline 1\,40\phantom{00} \\ -1\,28\phantom{00} \\ \hline 120\phantom{0} \\ -112\phantom{0} \\ \hline 80 \\ -80 \\ \hline 0 \end{array}$$

And then convert to a percentage: $0.1875 = 18.75\%$.

3. (a) $6$

$50\%$ of $12$ is $0.50 \times 12 = 6$

(b) $18$

$120\%$ of $15$ is $1.2 \times 15 = 18$

(c) $4.32$

$6\%$ of $72$ is $0.06 \times 72 = 4.32$

(d) $3$

$1.5\%$ of $200$ is $0.015 \times 200 = 3$

4. (a) $20\%$

Divide the part by the whole: $\frac{13}{65} = \frac{1}{5}$, which is the same as $0.2$ or $20\%$.

(b) $200\%$

Divide, keeping careful track of which number represents $100\%$:

$\frac{80}{40} = 2$, which is the same thing as $200\%$.

(c) $50\%$

Divide, keeping careful track of which number represents $100\%$:

$\frac{40}{80} = 0.5$, which is the same thing as $50\%$.

(d) $62.5\%$

Divide: $1.25 \div 2 = 0.625$, which is $62.5\%$.

5. (a) $1000$

Divide the part by the percentage. If $300$ is $30\%$, then $\frac{300}{30} = 10$ is $1\%$. Finally, $100\%$ is $10 \times 100 = 1000$. (Alternatively, you can just divide the part by the percentage converted to a decimal:

$\frac{300}{0.30} = 1000$.)

(b) $375$

Divide the part by the percentage. If $60$ is $16\%$, then $\frac{60}{16} = \frac{15}{4}$ is $1\%$,

which means that $100\%$ is $\frac{15}{4} \times 100 = \frac{15}{\cancel{4}} \times \overset{25}{\cancel{100}} = 375$.

(c) 30

Divide the part by the percentage expressed as a decimal. If $27$ is $90\%$, then $100\%$ is $\frac{27}{0.9} = \frac{270}{9} = 30.$

(d) 15

Divide the part by the percentage expressed as a decimal. If $36$ is $240\%$, then $100\%$ is $\frac{36}{2.40} = \frac{360}{24} = 15.$

6. $75\%$

$17$ out of $68$ is the same thing as $\frac{17}{68} = \frac{1}{4}$, or $25\%$. That's how much Mr. Schnitzel graded Tuesday night—leaving $100\% - 25\% = 75\%$ for later.

7. $18\%$

Dan has a total of $100\%$ of intellectual brain energy to spend. His other activities take up $37\% + 20\% + 16\% + 9\% = 82\%$ of his energy, which leaves $100\% - 82\% = 18\%$ for appreciating art.

8. 4 Labs

First, we find out how many dogs Laurie has: $70\%$ of $20$ pets means $0.7 \times 20 = 14$ parrots, which leaves $20 - 14 = 6$ dogs.

Now, we narrow our focus to dogs only. $33\frac{1}{3}\%$ is the same thing as $\frac{1}{3}$, so $33\frac{1}{3}\%$ of $6$ dogs is $2$ poodles. That leaves $6 - 2 = 4$ Labrador retrievers.

## CHAPTER 12

1. (a) $169$

$13 \times 13 = 169$

(b) $784$

$28 \times 28 = 784$

(c) $343$

$7 \times 7 \times 7 = 49 \times 7 = 343$

(d) $625$

$5 \times 5 \times 5 \times 5 = 25 \times 25 = 625$

2. (a) $6^9$

$$6^2 \times 6^7 = 6^{2+7} = 6^9$$

(b) $6^6$

$$6^{11} \div 6^{7-2} = 6^{11-(7-2)} = 6^6$$

(c) $6^8$

$$\frac{6^{25}}{6^{17}} = 6^{25-17} = 6^8$$

(d) $6^{63}$

$$(6^7)^9 = 6^{7 \times 9} = 6^{63}$$

(e) $6^5$

Recombine the powers of 2 and 3 into powers of 6:

$$2^5 \times 3^5 = (2 \times 3)^5 = 6^5$$

(f) $6^{14}$

Again, we need to recombine the powers of 2 and 3 into powers of 6. There are many different ways to do this, but here is one:

$\dfrac{27^{14} \times 2^{14}}{3^9 \times 3^{19}} = \dfrac{27^{14} \times 2^{14}}{3^{28}}$. Since there's a power of $2^{14}$ in the numerator, let's pull out a power of $6^{14}$ and hope that the rest cancels:

$$\frac{(27 \times 2)^{14}}{3^{28}} = \frac{(6 \times 9)^{14}}{3^{28}} =$$

$$= \frac{6^{14} \times 9^{14}}{3^{28}} = \frac{6^{14} \times (3^2)^{14}}{3^{28}}.$$

Luckily, $(3^2)^{14}$ is the same thing as $3^{28}$, so only $6^{14}$ is left.

3. (a) $4^6$

$2^{12} = 2^{2 \times 6} = (2^2)^6 = 4^6$

(b) $4^{10}$

$16^5 = (4^2)^5 = 4^{2 \times 5} = 4^{10}$

(c) $4^{24}$

$64^8 = (4^3)^8 = 4^{3 \times 8} = 4^{24}$

(d) $4^{33}$

Since $8$ is not a power of $4$, first convert everything to powers of $2$: $8^{22} = (2^3)^{22} = 2^{3 \times 22} = 2^{66}$. Now convert back into powers of $4$:

$2^{66} = 2^{2 \times 33} = (2^2)^{33} = 4^{33}$.

(e) $4^9$

First re-express $4^{10}$ in terms of $4^9$ by factoring: $4^{10} - 3 \times 4^9 = 4 \times 4^9 - 3 \times 4^9$. Next, use the distributive property backwards: $4 \times 4^9 - 3 \times 4^9 = (4 - 3) \times 4^9$, which simplifies to $1 \times 4^9$, or $4^9$.

(f) $4^4$

$2^7 + 2^7$ is the same thing as $2^7 \times 2$, which can be rewritten as $2^8$, or $(2^2)^4$.

4. (a) $2.25$

$1.5 \times 1.5 = 2.25$

(b) $0.81$

$0.9 \times 0.9 = 0.81$

(c) $0.0324$

$0.18 \times 0.18 = 0.0324$ because $18 \times 18 = 324$.

(d) $\frac{4}{25}$

$\frac{2}{5} \times \frac{2}{5} = \frac{4}{25}$

(e) $5\frac{4}{9}$ (equivalently, $\frac{49}{9}$)

$2\frac{1}{3} = \frac{7}{3}$, so $\left(2\frac{1}{3}\right)^2 = \frac{7}{3} \times \frac{7}{3} = \frac{49}{9}$, which is the same thing as $5\frac{4}{9}$.

5. (a) $8$

$10^7$ is a $1$ followed by seven 0s, for a total of eight digits.

(b) $96$

$2^4 + 3 = 19$. Moreover, $100000 = 10^5$, so $100000^{16+3} = \left(10^5\right)^{19}$, or $10^{95}$. That's a $1$ followed by ninety-five 0s, for a total of ninety-six digits.

## CHAPTER 13

1. (a) $1000$

$1{,}000{,}000$ has six 0s, so its square root must have three 0s: $1000^2 = 1{,}000{,}000$.

(b) $17$

Guess and check by hook or by crook. For example, $10^2 = 100$ and $20^2 = 400$, so $\sqrt{279}$ is between $10$ and $20$. Moreover, $15^2 = 225$, so $\sqrt{279}$ is $16$, $17$, $18$, or $19$. Since $279$ is odd, it can't be divisible by $16$ or $18$. So try both $17$ and $19$: $19 \times 19 = 361$, but $17 \times 17 = 289$.

(c) 24

Guess and check. For example, $20^2 = 400$ and $30^2 = 900$, so $\sqrt{576}$ is between 20 and 30. Moreover $25^2 = 625$, so $\sqrt{576}$ is less than 25—which leaves 21, 22, 23, and 24. And 576 is divisible both by 2 and 3, so 24 is a good guess. Alternatively, you can factor $576 = 2 \times 2 \times 2 \times 2 \times 2 \times 2 \times 3 \times 3 = 2^6 \times 3^2 = 64 \times 9$. Therefore $\sqrt{576} = \sqrt{64} \times \sqrt{9} = 8 \times 3$.

(d) 49

(e) $\frac{8}{3}$

$$\sqrt{\frac{64}{9}} = \frac{\sqrt{64}}{\sqrt{9}} = \frac{8}{3}$$

(f) 125

$$\sqrt{5^6} = \sqrt{(5 \times 5 \times 5) \times (5 \times 5 \times 5)} = 5 \times 5 \times 5 = 125$$

(g) 81

$$\sqrt{3^5 \times 3^3} = \sqrt{3^8} = \sqrt{3^4 \times 3^4} = 3^4 = 81$$

(h) 2.6

One way to think about this is to look for $\sqrt{\frac{676}{100}}$. Since 676 is between 625 and 900, $\sqrt{676}$ is between 25 and 30. Guess and check to determine that $26 \times 26 = 676$, which means that $\sqrt{\frac{676}{100}} = \frac{26}{10} = 2.6$.

(i) 4

We can try pulling the same trick we used in the last chapter. Since $2^5 = 2^4 \times 2$, or $2^4 + 2^4$, we can simplify $\sqrt{2^5 - 2^4} = \sqrt{2^4}$, which is the same thing as $\sqrt{2^2 \times 2^2}$, or $2^2$.

2. (a) 10

$6^2 + 8^2 = 36 + 64 = 100$, which is also $10^2$.

(b) 17

$8^2 + 15^2 = 64 + 225 = 289$, which is also $17^2$, from 1(b).

(c) 13

$5^2 + 12^2 = 25 + 144 = 169$, which is also $13^2$.

(d) 24

$25^2 - 7^2 = 625 - 49 = 576$, which is also $24^2$, from 1(c).

3. (a) 15

$3\sqrt{25} = 3 \times 5 = 15$

(b) $6\sqrt{2}$

$\sqrt{72} = \sqrt{2 \times 4 \times 9} = 2 \times 3\sqrt{2}$

(c) $\sqrt{115}$

$115 = 5 \times 23$, so there are no repeated factors and the expression is already simplified.

(d) $6\sqrt{7}$

$\sqrt{252} = \sqrt{4 \times 9 \times 7} = 2 \times 3\sqrt{7}$

(e) $8\sqrt{2}$

$\sqrt{128} = \sqrt{2 \times 64} = 8\sqrt{2}$

(f) $2^3 \times 3^4 \times 5 \times 7\sqrt{15}$

The thing to do is pull out even powers, which make perfect squares:

$$\sqrt{2^6 \times 3^9 \times 5^3 \times 7^2} = \sqrt{2^6 \times 3^8 \times 5^2 \times 7^2 \times 3 \times 5}$$
$$= 2^3 \times 3^4 \times 5 \times 7\sqrt{3 \times 5}.$$

4. (a) $1$

$\frac{6\sqrt{2}}{2\sqrt{18}} = \frac{6\sqrt{2}}{2\sqrt{2 \times 9}} = \frac{6\sqrt{2}}{2 \times 3\sqrt{2}} = \frac{\cancel{6}\cancel{\sqrt{2}}}{\cancel{2} \times \cancel{3}\cancel{\sqrt{2}}}$, so everything cancels.

(b) $\frac{\sqrt{7}}{7}$

$$\frac{1}{\sqrt{7}} = \frac{1}{\sqrt{7}} \times \frac{\sqrt{7}}{\sqrt{7}} = \frac{\sqrt{7}}{7}$$

(c) $\frac{3\sqrt{2}}{2}$

$$\frac{21}{\sqrt{98}} = \frac{21}{\sqrt{2 \times 49}} = \frac{\overset{3}{\cancel{21}}}{\cancel{7}\sqrt{2}} \times \frac{\sqrt{2}}{\sqrt{2}} = \frac{3\sqrt{2}}{2}$$

(d) $2\sqrt{3}$

$$\frac{4\sqrt{24}}{\sqrt{32}} = \frac{4\sqrt{4 \times 2 \times 3}}{\sqrt{16 \times 2}} = \frac{4 \times 2\sqrt{2 \times 3}}{4\sqrt{2}} = \frac{\cancel{4} \times 2\sqrt{\cancel{2} \times 3}}{\cancel{4}\cancel{\sqrt{2}}} = 2\sqrt{3}$$

(e) $3 + \frac{5}{3}\sqrt{3}$

$$\sqrt{27} + \sqrt{9} - \frac{4}{\sqrt{3}} = \sqrt{3 \times 9} + 3 - \frac{4}{\sqrt{3}} \times \frac{\sqrt{3}}{\sqrt{3}}$$
$$= 3\sqrt{3} + 3 - \frac{4}{3}\sqrt{3}$$

Since $3\sqrt{3}$ and $\frac{4}{\sqrt{3}}$ are like terms, they can be combined:

$3\sqrt{3} - \frac{4}{3}\sqrt{3} = \frac{9-4}{3}\sqrt{3} = \frac{5}{3}\sqrt{3}$. The final answer is

$3 + \frac{5}{3}\sqrt{3}$.

5. (a) 100

$1{,}000{,}000$ has six 0s, so its cube root should end in two.

$100^3 = 1{,}000{,}000$

(b) 6

Guess and check. For example, $10^3 = 1000$, so $\sqrt[3]{216}$ is less than $10$. Moreover, $216$ is even and divisible by $3$, so it makes sense to try $6$.

(c) 12

Guess and check. For example, $10^3 = 1000$ but $20^3 = 8000$, so $\sqrt[3]{1728}$ is less than $10$ and $20$. Since $1728$ is divisible by both $2$ and $3$, both $12$ and $18$ are good numbers to check. Alternatively, you can factor: $1728 = 2^6 \times 3^3$, so its cube root is $2^2 \times 3 = 12$.

(d) 4

From the table in the chapter, we know that $256 = 2^8$, which means that it's equal to $\left(2^2\right)^4$. That means that $\sqrt[4]{256} = 2^2 = 4$.

(e) 2

Guess and check—or look at the list of powers of $2$ in the chapter.

(f) 100

$$10^{12} = \left(10^2\right)^6, \text{ so } \sqrt[6]{10^{12}} = 10^2 = 100.$$

## CHAPTER 14

1. (a) different

$|-13| = 13$ is positive, whereas $-13$ is negative.

(b) same

$|4.5| = 4.5$

(c) same

$\left|-\frac{3}{4}\right| = \frac{3}{4}$

(d) same

Both are equal to $-67$.

(e) different

$-9$ is negative, whereas $9$ is positive.

(f) same

Both are equal to $0.07$.

(g) different

$-|-8| = -8$ is negative, whereas $8$ is positive.

(h) same

$-|-100| = -100$

(i) same

Both are equal to $0$.

(j) same

Both are equal to $1$.

(k) different

$|4 - 7| = 3$, whereas $|4 + 7| = 11$.

2. (a) $-1$

$8 - 9 = -1$

(b) $5$

$-2 + 7 = 5$

(c) $-8$

$-12 - (-4) = -12 + 4 = -8$

(d) $10$

$15 + (-5) = 15 - 5 = 10$

(e) $28$

$11 - (-17) = (11 + 17) = 28$

(f) $-0.75$

$-0.5 - 0.25 = -(0.5 + 0.25) = -0.75$

(g) $-\frac{5}{6}$

$-\frac{1}{2}+\left(-\frac{1}{3}\right)=-\left(\frac{1}{2}+\frac{1}{3}\right)=-\frac{5}{6}$

3. (a) 1

$(-5)\div(-5)=(5\div 5)=1$

(b) –7

$56\div(-8)=-(56\div 8)=-7$

(c) –21

$7\times(-3)=-(7\times 3)=-21$

(d) 48

$(-4)\times(-12)=4\times 12=48$

(e) –3

$-45\div 15=-(45\div 15)=-3$

(f) $\frac{1}{4}$

$-\left(-\frac{3}{4}\right)\div\frac{6}{2}=\frac{3}{4}\div\frac{6}{2}=\frac{3}{4}\div 3=\frac{1}{4}$

4. (a) 15

$-(2-7)\times|-2+5|=-(-5)\times|3|=5\times 3=15$

(b) 25

$|-9+4|^2=|-5|^2=5^2=25$

(c) –4

$\frac{3|11-7|-4}{-|7-11|+2}=\frac{3|4|-4}{-|-4|+2}=\frac{3\times 4-4}{-4+2}=\frac{12-4}{-2}=-\frac{8}{2}=-4$

(d) 6

$-2|4-|5-9||+6=-2|4-|-4||+6=$

$-2|4-4|+6=-2|0|+6=6$

## CHAPTER 15

1. (a) $550 : 220 = 5 : 2$

   (b) $72 : 56 = 9 : 7$

   (c) $3 : 5$

   I don't know $375$ out of $600$ words, which means I *do* know $600 - 375 = 225$ words. The ratio of words I know to words I don't know is $225 : 375$. Divide each side by $75$ to simplify.

2. (a) $50$

   $\frac{2}{5} = \frac{10}{?}$, so $2 \times (?) = 50$.

   (b) $24$

   Set up the proportion: $\frac{8}{28} = \frac{?}{84}$. Reduce: $\frac{\overset{2}{\cancel{8}}}{\underset{7}{\cancel{28}}} = \frac{?}{84}$, so

   $2 \times 84 = (?) \times 7$.

   (c) $21$

   Set up the proportion: $\frac{9}{?} = \frac{75}{175}$. Again, reduce: $\frac{9}{?} = \frac{\overset{3}{\cancel{75}}}{\underset{7}{\cancel{175}}}$, so

   $9 \times 7 = 3 \times (?)$.

   (d) $1$

   Set up the proportion: $\frac{6}{4} = \frac{1.5}{?}$, so $6 \times ? = 1.5 \times 4$, or $6 \times ? = 6$.

3. (a) $30$ inches

   Set up the proportion: $\frac{5}{6} = \frac{25}{\text{leg length}}$, so

   $5 \times (\text{leg length}) = 25 \times 6$, and $\text{leg length} = \frac{25 \times 6}{5} = 30$.

(b) About 38 games

We know the ratio of games won to games played as well as the number of games played, so we can first find out the number of games *won*.

Set up the proportion: $\frac{5}{8} = \frac{\text{games won}}{100}$,

so $5 \times 100 = (\text{games won}) \times 8$, and

games won $= \frac{5 \times 100}{8} = 62.5$. But what to do with a fractional number of games? Well, this ratio is *on average*, and we're asked for an approximate number of games lost. $62.5$ games won means $100 - 62.5 = 37.5$ games lost. Round up or down.

(c) 4 packets

Set up the proportion: $\frac{6 \text{ packets}}{2\frac{1}{4} \text{cups}} = \frac{? \text{ packets}}{1\frac{1}{2} \text{cups}}$.

Cross-multiply:

$6 \times 1\frac{1}{2} = ? \times 2\frac{1}{4}$, or $6 \times \frac{3}{2} = ? \times \frac{9}{4}$, or $9 = ? \times \frac{9}{4}$,

so ? = 4.

(d) $48$ bushels; $25$ bunnies

Set up the proportions. First, $\frac{36 \text{ bushels}}{15 \text{ bunnies}} = \frac{? \text{ bushels}}{20 \text{ bunnies}}$.

So $? = \frac{36 \times 20}{15} = 48.$

Next, $\frac{36 \text{ bushels}}{15 \text{ bunnies}} = \frac{60 \text{ bushels}}{? \text{ bunnies}}$, so $? = \frac{60 \times 15}{36} = 25.$

4. (a) $2 : 4 : 5$

We can reduce by dividing each element by 4, which is the GCF of $8$, $16$, and $20$.

(b) 12 pieces of dill-flavored taffy; 15 pieces of caramel-popcorn taffy
Since the ratio is the same, $2 : 4 : 5 = 6$ : dill : caramel-popcorn. Although we can't rewrite this equivalence as a neat fraction, we can break it up into two proportions: $2 : 4 = 6$ : dill and $2 : 5 = 6$ : caramel-popcorn. Solve both:

$$\text{dill:} \quad \frac{2}{4} = \frac{6}{\text{dill}}$$

so dill $= \frac{6 \times 4}{2} = 12.$

$$\text{caramel-popcorn: } \frac{2}{5} = \frac{6}{\text{caramel-popcorn}},$$

so caramel-popcorn: $= \frac{6 \times 5}{2} = 15.$

5. 21 Mexican cousins; 9 Icelandic cousins
We know the ratio of Mexican cousins to Icelandic cousins, but it would be great to get the ratio of, say, Mexican cousins to total cousins. Then we can use the fact that Emily P. has 30 cousins to get the exact number of Mexican cousins.

Since Emily P. has 7 Mexican cousins for every 3 Icelandic cousins, she has 7 Mexican cousins for every 10 total cousins. (This total comprises the 7 Mexican cousins and 3 Icelandic cousins.) So

$\frac{\text{Mexican cousins}}{\text{all cousins}} = \frac{7}{10}$. Since (all cousins) = 30,

we have $\frac{\text{Mexican cousins}}{30} = \frac{7}{10}$. Therefore

Mexican cousins = $\frac{7 \times 30}{10} = 21$, which leaves $30 - 21 = 9$ Icelandic

cousins.

# CHAPTER 16

1. (a) 10

Put the numbers in order: 3, 3, 3, 8, 9, 10, 54, 67, 89, 89, 100. The median is the middle number (sixth of eleven): 10.

(b) 39.55

The mean is the sum divided by the number of numbers (eleven):

$$\frac{3+3+3+8+9+10+54+67+89+89+100}{11} = \frac{435}{11} = 39.54545...$$

Rounded to the nearest hundredth, 39.54545... becomes 39.55.

(c) Yes, the mode is 3.

There are two 89s and three 3s; 3 is the most frequently occurring number.

(d) Mean increases to 44.67; median increases to 32; mode stays 3. Adding a large number to the list increases the mean.

$$\frac{3+3+3+8+9+10+54+67+89+89+100+101}{12} = \frac{435}{12} = 44.67...$$

The median also changes. Since there are now an even number of entries, the median is the average of the middlemost two (sixth and seventh of twelve): $\frac{10+54}{2} = 32$.

The mode doesn't change; it's still 3.

2. 45

The average of six numbers is their sum divided by 6:

$$\frac{46+50+34+20+69+?}{6} = 44.$$

We can cross-multiply to get $46 + 50 + 34 + 20 + 69 + ? = 44 \times 6 = 264$. So the sum of all six numbers is 264, and we can find the mystery number by subtracting off everything else:

$$264 - 46 - 50 - 34 - 20 - 69 = 45.$$

3. (a) $38.5\%$

The percentage is given by $\frac{\text{part}}{\text{whole}} = \frac{\$500}{\$1300} = 0.385$. Move the decimal point two digits to the right to convert from a decimal to percentages.

(b) Microphone

From the pie chart, the only expense that takes up about a quarter-circle is the microphone. Check: One-fourth of $\$1300$ is $\$325$, which is close to the $\$300$ spent on the mike.

(c) Party; more

From the pie chart, the expense that is closest to the profit slice is the party. The party slice is a little bigger.

(d) Parking ticket and guitar picks; less

From the pie chart, the two expenses that, put together, make something close to the profit slice, are the parking ticket and the guitar picks. Their share together is a little smaller than the profit slice.

(e) About $\$165$

The profit slice is less than the cost of the party ($\$200$) but more than the cost of the picks and the parking ticket ($\$80 + \$50 = \$130$). Estimate something in the middle.

(f) $\$170$ to split; Kyle gets $\$42.50$

Subtract all the other expenses from the total:

$$\$1300 - \$300 - \$200 - \$80 - \$50 = \$170.$$

Splitting $\$170$ four ways gives $\$170 \div 4 = \$42.50$ for Kyle.

4. (a) $12$ on Wednesday; $4$ on Thursday

There are $4$ pictograms for Wednesday, and each one represents $3$ tripping incidents, for a total of $3 \times 4$ trips.

There is a bit more than one pictogram for Thursday, so I tripped $4$ or $5$ times. It looks like the partial pictogram is about a third (compare with the partial on Friday), so $4$ trips is the best guess.

(b) Tuesday and Thursday

Fewer than $6$ times corresponds to fewer than $2$ pictograms. That's Tuesday and Thursday.

(c) Around 7 or 8 times on average

It looks like the average will be between 2 and 3 pictograms, which corresponds to 7 or 8 trips.

(d) 7.7 times

I tripped 6 times on Monday, 3 times on Tuesday, 12 times on Wednesday, about 4 times on Thursday, about 14 times on Friday, 6 times on Saturday, and 9 times on Sunday. The weekly average is

$$\frac{6+3+12+4+14+6+9}{7} = 7.7 \text{ trips.}$$

5. (a)

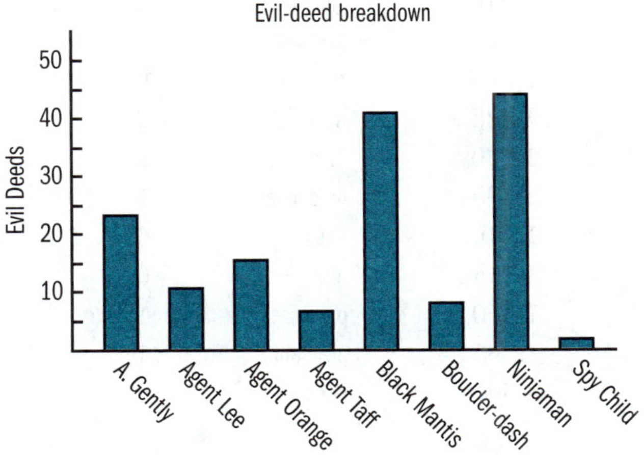


From the graph, we see that the average—the number that balances the highs and lows—is around 20.

(b) 152 evil deeds total

Add everything up:

$$23 + 11 + 16 + 7 + 41 + 8 + 44 + 2 = 152.$$

(c) 19 evil deeds on average

Divide the sum by the number of spies: $\frac{152}{8} = 19.$

## REVIEW PROBLEM

(a) Glorious at $12{:}15$; mediocre at $12{:}35$

(b) $12{:}05$, $12{:}50$, and $12{:}55$

(c) Roughly from $12{:}15$ to $12{:}25$

(d) Average is $\frac{1}{4}$, a little better than fine.

First, make a table of Rebecca's mood states:

| Time | State | Point |
|---|---|---|
| 12:05 | pleasant | 1 |
| 12:10 | glorious | 2 |
| 12:15 | glorious | 2 |
| 12:20 | fine | 0 |
| 12:25 | dreadful | –2 |
| 12:30 | mediocre | –1 |
| 12:35 | mediocre | –1 |
| 12:40 | fine | 0 |
| 12:45 | fine | 0 |
| 12:50 | pleasant | 1 |
| 12:55 | pleasant | 1 |
| 1:00 | fine | 0 |

Take the average:

$$\frac{1+2+2+0-2-1-1+0+0+1+1+0}{12} = \frac{3}{12} = \frac{1}{4}.$$

So on average, Rebecca felt just a bit better than fine.

(e) $25\%$

Rebecca felt fine $15$ minutes out of $60$, which is $\frac{15}{60} = 0.25$ of her time, or $25\%$.

(f) **Mood state** | **Fraction**

| **Mood state** | **Fraction** |
| --- | --- |
| Glorious | $\frac{10}{60} = \frac{1}{6}$ |
| Pleasant | $\frac{20}{60} = \frac{1}{3}$ |
| Fine | $\frac{15}{60} = \frac{1}{4}$ |
| Mediocre | $\frac{10}{60} = \frac{1}{6}$ |
| Dreadful | $\frac{5}{60} = \frac{1}{12}$ |

Sum is $\frac{1}{6} + \frac{1}{3} + \frac{1}{4} + \frac{1}{6} + \frac{1}{12} = \frac{2}{12} + \frac{4}{12} + \frac{3}{12} + \frac{2}{12} + \frac{1}{12} = 1$

(g) **Rebecca's moods**

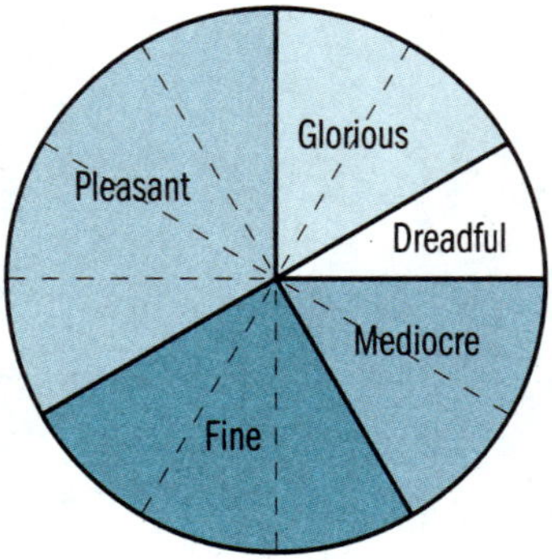

# EASY GUIDE TO ALGEBRA

# CONTENTS

# 1

# CHAPTER 1
## THE BASICS: MEET LETTER X

Okay, I know you're probably not too interested in math. But no worries–I'm here to help. I suppose I should introduce myself. I'm Ryan Hall, algebra tutor *extraordinaire*. Yes, get a life, I know. I've been talking to you for all of two seconds and you probably have me pretty much figured out: All I know is algebra. I literally *have no life*. But this is not entirely my fault. You see, I was in this skateboarding accident, and I hit my head, and now I have no ability to form new memories. I can remember everything that happened before my accident, but my short-term memory is completely gone. Stop me if I've explained this to you already. I don't remember seeing you before in my life, but even if I had met you, I wouldn't remember.

I bet the mere mention of algebra is enough to give you a bit of a sweaty-palm feeling, right? I'm here to tell you, "Nah, it's not a problem."

## WHAT IS ALGEBRA?

When you get right down to it, **algebra** is finding an unknown number in a problem. That's it. The unknown number is referred to by a letter, such as an *x*.

What do I mean by an unknown number? For instance, you know that $2 + 1 = 3$. But what if you took out the 1 and put in an *x*? Suddenly, you've got $2 + x = 3$. Now you're doing algebra! The *x* is an unknown number.

Of course, most problems are more difficult than that, but right now we're just talking about the concept, okay? When you do algebra, you want to know what number to put in place of that *x* in order to solve the problem.

# VARIABLES

Before we get too far, let's get some vocabulary out of the way. I'll bet you didn't think there would be vocabulary to learn in an algebra class. Don't worry, there's not too much.

Probably the most important word to know is **variable**. That's the name math folks give that unknown number we were just talking about. That's the letter $x$ in the problem we looked at above. It's called the variable. In this problem:

$$4 + x = 12$$

$x$ is the variable. It could also be a $y$ ($4 + y = 12$), or it could be an $n$ ($4 + n = 12$). In all three cases, the answer, 8, is represented by a variable: $x$, $y$, $n$. Get it? It's any letter that represents a number.

But a variable has no fixed number value. It changes from problem to problem. That's right: it *varies*. Like, suppose we change the 12 in this example to 20:

$$4 + x = 20$$

Now the value of $x$ (or $y$, or $n$) changes to 16.

# EXPRESSIONS

Just a couple more vocabulary words and we'll be free and clear. In algebra, an **expression** is a group of letters and numbers together, connected with symbols such as a plus or minus sign. Here is an example of an expression in algebra:

$$4x + 10$$

You can break up an expression into smaller parts. The 10 and the $4x$ are both called **terms**. They are individual parts of an expression. The 4 in this example is called the **coefficient**. A coefficient is a number multiplied with a variable.

**term**

$$\overbrace{4x + 10}$$

**variable**

**coefficient**

# TRANSLATING ALGEBRA INTO ENGLISH

One problem people have when they do algebra is that they see all these numbers and they freak out. But algebra can be expressed in plain English. Watch:

$$3x - 7$$

This means "3 times a number minus 7." Here's another one:

$$8 + 5x$$

This can be translated into plain English by saying "8 plus 5 times a number."

# TRANSLATING ENGLISH INTO ALGEBRA

You probably never thought you would have to, but you can also translate English into algebraic expressions. Use $x$ whenever you're talking about "a number."

So, when you say, "a number minus 4," you are saying,

$$x - 4$$

The point of this is not for you to run out and write algebra problems. The point is for you to understand that you can make sense out of an algebraic expression and not be intimidated by it. We'll get into how to solve the expressions, but you have to start with understanding what the expression means.

Sometimes, you need to deal with two different variables in the same expression. They must be stated as two different letters. So, when you say, "12 plus a number minus another number," you express it as

$$12 + x - y$$

# AVOIDING CONFUSION

Here's an interesting problem with algebra:

$$17 \times x$$

The multiplication sign looks an awful lot like the variable, doesn't it? People thought there ought to be a way to avoid this confusion. There are three different ways you can write that same expression without confusion.

$$17 \cdot x$$
$$17x$$
$$17(x)$$

Each means the same thing: "17 times a number." The symbols are all used to express multiplication.

# YOUR TURN

1. Which is the *variable*?
   $7y - 11$

2. Is this a *term* or an *expression*?
   $8 + 14 - x$

3. Which is the *coefficient*?
   $19n \cdot y$

Write these expressions:

4. 12 times a number
5. A number plus another number
6. 9 plus a number times 8
7. 8 times a number minus that same number
8. A number minus 1 plus another number
9. 3 times a number plus 5 times another number
10. 450 times 11 plus 2 times a number

Write what these expressions mean in English:

11. $4y + 0$

12. $81 \cdot x - y$

13. $2x + 3y$

14. $7 + y$

15. $12n - 3c$

# 2

# CHAPTER 2
## POSITIVE AND NEGATIVE NUMBERS



Now, I know you're probably not too interested in math. Very few of us are. But just because you don't like math doesn't mean you get to skip it for the rest of your life. I suppose I should introduce myself. My name is Ryan Hall, and I'm here to help. One thing you should know about me: I injured my head in an accident, and now I can't make new memories. I don't remember yesterday or the day before any more than if they were tomorrow.

Please excuse me if I seem a little tense today. I'm about to give a tutoring session to this guy named Craig—the star player of the basketball team. Don't get me wrong. I love the people I tutor—I practically live for them. But Craig's a big shot, you know? Big man on campus. I bet he thinks that just because he doesn't like math, he gets to skip it for the rest of his life. Well, I've got news for him.

But that's not why I'm nervous. What's making me feel weird is that I actually used to be friends with Craig back in sixth grade. We used to spend hours in his driveway, doing ollies and kick flips on our skateboards all afternoon, until his mother would call us in for some chocolate chip cookies and milk. His mother—don't get me started on *her*. But all that was a long time ago—I haven't talked to Craig since we were little kids. At least, I don't think I have.

Just seeing Craig's driveway is enough to send me flying down memory lane. And what have we here? I might have known. There in the driveway, his back still turned to me, is Craig. On his skateboard. He looks *exactly* the same—same baggy jeans, same Stussy hooded sweatshirt. He's about to do a kick flip. A devilish impulse comes over me. I creep up behind him and knock him off his board as he's doing the flip.

"Yo, Craig! How *are* you, dog? How you been?"

He starts to cry.

And then I realize: if this were really Craig Gurevich, *he would have grown up too*.

Whoa. Okay, so I just knocked over an innocent sixth-grader. Not Craig. Sometimes I get a bit confused about things, I admit. But *algebra*? No worries there. I'm here to help.

# SIGNED NUMBERS



Remember number lines? A **number line** is a line that represents numbers in a sequence. The most interesting thing about number lines is that they continue forever in both directions. So, when you have a 0 on a number line, you have positive numbers on the right side and negative numbers counting down on the left.

$$-3 \quad -2 \quad -1 \quad 0 \quad 1 \quad 2 \quad 3$$

A **signed number** is a number with a plus or a minus sign in front of it. You usually won't see a plus sign to indicate a positive number, such as +3. When the number is written as just 3, it's understood that it's a positive number.

But for a negative number, we have to indicate that it's negative with a minus sign, such as -45.

## ABSOLUTE VALUE

Whether you're talking about +37 or -37, you're still talking about 37 units of something, right? Whether it's 37 units to the right on the number line or 37 units to the left on the number line, the value of the number is still 37.

That's why the number that remains when you remove a sign from a number is called the **absolute value** of the number. The symbol | | is used to indicate absolute value. The absolute value of -10 is 10, and it's written as

$$|-10| = 10$$

Remember, the absolute value of a number is always a positive number. It's the number without a sign in front of it.

One word of advice about signed numbers. They're very important in algebra. A negative sign is often a minus sign, a positive sign is often an addition symbol, and signs change so often in algebra from positive to negative and back again that your head will spin if you're not ready for it. So here we go.

## ADDING SIGNED NUMBERS

To add two signed numbers together, first take a look at the signs. If they're both the same sign, add the numbers together (the absolute value of the numbers) and then just tack on the sign when you're finished. In effect you're just counting backward on the number line. Like this:

$$(-9) + (-13) = -22$$

Notice that we put the negative numbers in parentheses so the plus and minus signs aren't confusing to read when we do operations. It makes sense, and it makes things clearer.

Now let's look at an example of adding two numbers with different signs:

$$8 + (-15)$$

Since one of the signs is negative, you subtract the smaller absolute value from the larger one. Then tack on the sign that belongs to the larger absolute value:

$$15 - 8 = 7$$

The answer becomes a negative number because it takes on the sign of the 15, which is negative:

$$8 + (-15) = -7$$

## SUBTRACTING SIGNED NUMBERS

To subtract signed numbers, add the opposite of the sign. Let's take a look at an example:

$$8 - (-2)$$

The opposite of -2 is 2. So, 8 + 2 = 10.

$$8 - (-2) = 10$$

Now try (-3) - (-5). The opposite of -5 is 5. So, -3 + 5 = 2. To do the subtraction problem, you have to remember how to do the addition problem. Subtract the absolute value of 3 from 5, and keep the sign positive because the sign on the 5 is positive.

## MULTIPLYING SIGNED NUMBERS



To multiply two signed numbers, multiply the absolute values of the numbers. If the signs are the same, it is positive. If the signs are not the same, make the answer negative. For example,

$$3(-11) = -33$$

First, multiply: 3 times 11 is 33. Then tack on the negative sign. If the signs are different, the answer is negative.

What if the signs are the same?

$$(-5)(-2) = 10$$

Multiply: 5 times 2 is 10. Since the signs are both negative, the answer is positive. If you are multiplying two numbers with the same sign, whether they've both negative or both positive, the answer will be positive.

## DIVIDING SIGNED NUMBERS

Dividing two signed numbers is a lot like multiplying them. Divide the absolute values for the numbers. If the signs are not the same, make the answer negative:

$$120 \div (-3) = -40$$

If the signs are the same, make the answer positive:

$$-35 \div (-7) = 5$$

### THE FATHER OF ALGEBRA

A Greek mathematician named Diaphanous, who lived around 250 BCE, is thought to be the first person to write down ways to express algebraic equations and solutions. While algebra had technically been around for about 3,000 years before Diaphanous, he was known as the first person to write this stuff down. He wrote a famous work called *Arithmetica*, which contains more than 150 algebra equations and solutions. But he didn't use the rules we use today for solving problems. In fact, he solved all his problems with different methods! And here's another kicker. He didn't use symbols, so every problem was expressed in words.

# YOUR TURN

1. Find | –2 |

2. Find | 0 |

3. Find | +7 |

4. Write the absolute value of -142 as an algebraic term.

5. Find | –40 |

Solve the following expressions:

6. -8 + 4

7. 14 + (-6)

8. (-5) - (-2)

9. (-8) - (+7)

10. (+40) - (-30)

11. -8(-4)

12. (-14)(-3)

13. (-5)(+9)

14. (-11)(15)

15. -12 ÷ 2

16. -20 ÷ -4

17. (-8) + (-9)

18. (-14) + (3)

19. 5 – (17)

20. (-4)(+10)

# 3

# CHAPTER 3
## SIMPLIFYING ALGEBRAIC EXPRESSIONS



Before I teach you how to simplify algebraic expressions, I should mention that since my accident, I can't form new memories at all. I don't remember whether I've met you before or what we talked about yesterday.

My name's Ryan Hall, by the way. I apologize if I seem a little tense today. I'm on my way to teach simplifying expressions to this basketball player named Craig. I'm meeting him at his house, and I haven't been there since sixth grade, when I embarrassed myself outrageously with his mom in this horrible scene that I'd rather not relive right now. Let's just say that I opened the wrong door and caught his mom in an embarrassing situation. So, I'm a little nervous.

Craig lets me in, and as I enter his house, I see that everything looks just as it did when we were in sixth grade—indeed, just as it did that fateful day as I fled down those stairs and out the door burning with shame and anguish.

But before I get too deep into nostalgia, who should enter the room but Craig's mother! She's smartly dressed in a white silk blouse with matching long, white silk scarf, her lustrous brown hair gathered back into a high ponytail, her pert little nose wrinkling as she smiles at us over the silver tray she's holding.

"How nice to see you, Ryan. I brought you boys some milk and cookies."

"Um, no thanks. They look great, but for some reason, I feel really full, Mrs. G."

"That's fine, Ryan."

What is wrong with me! The smile is gone. Does she think I was rude? Why didn't you take the cookie, you idiot! Say something—save the moment—say anything!

"Uh, Mrs. G., you haven't changed at all. You're just as beautiful as you were when I used to come here."

*Argh*. Some days, it's just easier to think about algebra. Got that sweaty-palm feeling yet? I'm here to tell you, "Nah, it's not a problem."

# PARENTHESES

Have you ever done math problems with parentheses? Maybe you've heard the mnemonic device for the **order of operations**. It goes Please Excuse My Dear Aunt Sally. That stands for Parentheses, Exponents, Multiplication, Division, and Subtraction. This tells you that you should work on expressions that are inside the parentheses first.

So, for example, when you see a problem like this, you know you should work on the terms in the parentheses first.

$$3 + 4 - (9 + x)$$

# BRACKETS AND BRACES

The parenthesis is not the only grouping symbol in the math world. Sometimes in algebra you'll find symbols called **brackets**. They look like this:

[ ]

And sometimes you'll find grouping symbols called **braces**. They look like this:

{ }

The symbols separate different sets of expressions. They can also be used along with parentheses in the same expressions. This allows you to have a long string of terms, with brackets inside parentheses inside braces. Don't worry. You'll see how all of this stuff looks when we get into solving equations. For now, it's just a heads-up. The symbols are meant to help, not to make things look more confusing.

# DISTRIBUTIVE PROPERTY

When you see expressions that are grouped, your first job is to ungroup them, or get rid of the parentheses. There's a little rule to help you do that. The rule is called the distributive property, and it says

$$a(b + c) = ab + ac$$

See how the *a* was "distributed" evenly to each letter across the parentheses? The *a* is a coefficient in this example. The *a* gets multiplied with each letter across the parentheses.

Let's try using the distributive property with a real problem:

$$7(x + 3)$$

Look! It's our first algebra problem with a variable! Call in the media. We're on our way. Let's distribute the 7 across the problem by multiplying it with each term:

$$7x + 7(3)$$

Okay, now take that one step further and do any other multiplication you can. We can't do $7x$ because we don't know what number $x$ represents. But we can finish up the 7(3):

$$7x + 21$$

The distributive property works the same way with subtraction. The rule is

$$a(b - c) = ab - ac$$

SIMPLIFYING EXPRESSIONS

# COMBINING "LIKE" TERMS

Now that you know how to remove grouping symbols by using the distributive property, you can work on longer problems. Just keep working through the line, one grouping at a time, from left to right:

$$3(x-1)-4(2-1x)-(x+1)$$

The minus signs between each term should be combined with the coefficients and treated as signed numbers. So, use your knowledge of signed numbers to work through each term. Notice that with the last grouping of numbers, you're distributing the negative sign across the parentheses. This changes the signs within the group, so the next step looks like this:

$$3x-3-8+4x-x-1$$

As you can see, when you use the distributive property, you are left with some terms that have variables and some that don't. You want to combine all of the "like" terms. That means you should group all of the variables together and all of the coefficients together. Your knowledge of signed numbers will help you here too:

$$(3x+4x-x)-3-8-1$$

With each step, you are combining another term. Keep going until the expression has just one variable, one coefficient, and one integer.

$$6x-12$$

# WORKING FROM THE INSIDE OUT

When you have a long expression with several grouping symbols, such as brackets and parentheses, work from the inside out. First, solve the smaller groups inside the large grouping symbol. Let's take a look at how that works:

| | |
|---|---|
| $3\{x - 2[5 - (x + 1)] + 1 - 2\}$ | Get rid of inside parentheses. |
| $3\{x - 2[5 - x - 1] + 1 - 2\}$ | Then get rid of brackets. |
| $3\{x - 10 + 2x + 2 + 1 - 2\}$ | Distribute the 3 across. |
| $3x - 30 + 6x + 6 + 3 - 6$ | Combine the like terms. |
| $9x - 27$ | |

# POLYNOMIALS, ETC.

Here's another word about vocabulary. Sometimes, you'll run across big, fancy words that mean simple, little things. *Polynomials, binomials, trinomials . . .* – these are all fancy words for the number of terms in an expression. Remember, terms are separated from other terms with a plus or minus sign.

A **binomial** is an expression with exactly two terms.

$3(4 + 2x) + 11$ binomial

A **polynomial** has more than two terms.

$5x + 6(2x) - 17 + (3 \cdot 4x)$ polynomial

A **trinomial** has exactly three terms.

$10y + 7y + 8$ trinomial (also a polynomial)

## BRACES OR BRACKETS?

What's the difference between parentheses, braces, and brackets? When should you use them? Well, there's really no difference at all except in the way they look. Just imagine a long equation with grouping symbols inside grouping symbols—and suppose only parentheses were used? It would look so confusing, you wouldn't know which group went with which parentheses. Adding brackets and braces makes it easier to tell groups apart. The order goes like this: { [ ( ) ] }. Parentheses are the most commonly used grouping symbol. Brackets are used to enclose a group with parentheses inside it. Braces enclose brackets and are the least used of the grouping symbols.

# YOUR TURN

Use the distributive property to simplify these expressions.

1. $14(x + y)$

2. $5(x - 4) + 4(3x + 7)$

3. $x - [7 - 2(x + 1)]$

4. $5(7x - 3 + x) - (x + 5)$

5. $-3(4 + x) - 5[x - (3 - x)]$

6. $y(8 + 2) + x(4 - 1)$

7. $3y(4 - 2y) - 2[y + (3x + 2x)]$

8. $2\{y - 3(6 - [2 + y]) + 3\} - 2$

9. $4 + 2(x - 2[5 + x])$

10. $14(2x + 3x) - x(5 - 4x)$

# 4

# CHAPTER 4
## SOLVING EQUATIONS

Hello! I'm Ryan, algebra tutor *extraordinaire*. Yes, get a life, I know. But listen—just because you don't like math doesn't mean you get to skip it for the rest of your life. I'm here to help.

One thing you should know about me is that I have no short-term memory. I can't form new memories at all. I remember everything up to my accident, but since then, it's a blank. But don't worry—I'm a whiz at this algebra stuff.

I admit I'm a little nervous today because I'm going to this guy Craig's house to tutor him for the first time. He and I used to be friends, but that was ages ago, when we were kids. I haven't been there since I embarrassed myself horribly in front of his mother back in sixth grade. I'm sure she's forgiven me, but I know she couldn't have forgotten, so I feel a bit awkward. My palms are a bit sweaty, to tell you the truth.

Just walking to Craig's house is enough to send me flying down memory lane. I turn into his driveway and am hit with a hundred flashbacks of the two of us skateboarding there and nearly breaking our necks daring each other to do stupid tricks. He was always more athletic than I was, but I wasn't too bad on a skateboard. Until my accident, of course.

Anyway, after Craig lets me in, it isn't long before I have to confront my worst fear. In from the kitchen, carrying a silver tray of milk and cookies, walks his mom. She's dressed from head to toe in winter white—a pashmina with a fringe and a long, tailored skirt with a slit up the side—and her lustrous brown hair is gathered up into a French twist.

"How nice to see you, Ryan. I brought you boys some milk and cookies."

I reach out to take a cookie. I want to take it. I really do. But somehow, the thought makes me slightly queasy. Why do I feel so full? *Disgustingly* full. "Er, no thanks, Mrs. G."

Idiot! Say something. Say something. Rescue the moment before she goes away.

"Um, Mrs. G.?" My voice cracks. "You look just as beautiful as you did when

I was a kid."

"Poor Ryan. You never change, do you?"

I can't remember if I change or not—I'd rather just think about algebra. Does the mere thought, the merest hint, of the word *algebra* make your skin crawl, your palms itch, and your breath come in short, ragged gasps? I bet it does. But I'm here to tell you, "Nah. It's not a problem."

# EQUATIONS

Have you noticed that up until now we haven't been working with an equal sign in any of our problems? That's not a coincidence. I did that on purpose. We've been working only with expressions so far. Now we're going to start working with equations. An equation is made up of two expressions that are equal. In other words, there's an expression, an equal sign, and another expression on the other side of it. Here's an example:

$$5(x-1) = 24$$

Let's learn how to deal with the left and the right side of the equal sign at the same time.

# FIRST-DEGREE EQUATIONS

In this chapter, we solve only first-degree equations. Here's what a first-degree equation is:

1. There's only one variable.
2. There are no exponents in the problem.
3. The variable is not in the denominator of a fraction.
4. The equation uses more than one operation.

This will keep things fairly simple for now until we feel comfortable doing what needs to be done to solve an equation and finally figure out what that pesky $x$ equals.

# ISOLATING THE VARIABLE

In order to solve an equation in algebra, you have to simplify the expressions as best as you can and then move some things around. What you are trying to do is put the $x$ all by itself on one side of the equation. That means that by the end of your problem, you'll end up with something that looks something like this:

$$x = 14$$

Remember, this is our end goal. It's called isolating the variable. There's a four-step plan for doing this that you can take with you forever. It's the basis for solving any algebra problem. Let's take one example problem and go through our four-step plan for solving equations. The first step is to *simplify*. Here we go.

## STEP 1: SIMPLIFYING

You already know how to do this by using the distributive property and combining like terms. This time, you do it for both sides of the equation. Here's our example:

$$2[3 + x + 4(2x - 1)] = 8 - x + 3$$

Work slowly and carefully, following the **order of operations**. That means that you work on parentheses first. If you see more than one kind of grouping symbol, solve for the innermost group first and work your way out. Luckily, on this equation, the right side of the equal sign doesn't need much simplifying, so let's concentrate on the left side:

| | |
|---|---|
| Simplify the innermost group. | $2[3 + x + 8x - 4] = 8 - x + 3$ |
| Distribute the 2. | $6 + 2x + 16x - 8 = 8 - x + 3$ |
| Combine like terms. | $-2 + 18x = 11 - x$ |

This is as far as you can go on each side of the equal sign because you've combined the like terms. All of the coefficients and variables are together, and all of the like terms are together. Step 1 is complete.

SOLVING EQUATIONS

## STEP 2: TRANSPOSING

Our next step is to **transpose**, or move, terms from one side of the equation to the other. You want to put all of the variables with their coefficients on one side of the equation and all of the terms without variables on the other side of the equation.

This transposing stuff works because you're really adding a 2 to both sides of the equation. When you do the same thing to both sides of the equation, everything stays equal, and that's what we want to do. So, on the left side of the equation, the 2 goes away because -2 plus 2 is 0. On the right side of the equation, you have your integers grouped together like you wanted. Both sides of the equation are treated equally, so everyone's happy.

You don't have to write this step out, because it gets a bit confusing. But there's one important thing to know when you transpose. You have to change the sign when you move something from one side of the equal sign to another. For example, if you have a positive number on the left expression and you want to move it to the right, you have to make it a negative number. Once it jumps over that equal sign, it becomes the opposite of itself.

Here's where we last left our problem:

$$-2 + 18x = 11 - x$$

Now, let's transpose.

| | |
|---|---|
| Move -2 to the right side, change signs. | $18x = 11 - x + 2$ |
| Move $-x$ to the left side, change signs. | $18x + x = 11 + 2$ |

## STEP 3: SIMPLIFYING AGAIN

Now that we've got all of the variables on one side of the equation, we can simplify again. Combine terms on both sides of the equation. When you're working with one side of the equation, don't forget that you have to do the same type of thing to the other side of the equation. So, if you're simplifying the coefficients on the left, simplify the terms on the right as well. You can do it in two steps to be clearer, but don't forget!

Here's where we left off:

| | |
|---|---|
| | $18x + x = 11 + 2$ |
| Combine like terms on the left. | $19x = 11 + 2$ |
| Combine like terms on the right. | $19x = 13$ |

## STEP 4: DIVIDING BY THE COEFFICIENT

The last step is to get that $x$ all by itself, without its coefficient. Remember how we had to change the sign on positive and negative numbers when we moved them to the other side of the equation? That's because we were changing a plus to a minus. They are inverse, or opposite, operations.

We have to do something similar with the coefficient. Since $19x$ really means 19 times $x$, we have to do the opposite operation when we move the 19 to the other side, which means we have to divide the other side of the equation by 19.

That means we have to make a fraction. The coefficient will become the denominator of the fraction, but you should probably show your work by multiplying each side by $^{19}/_{19}$.

So, the last step of our problem, $19x = 13$, changes to

$$\frac{\cancel{19}x}{\cancel{19}} = \frac{13}{19}$$

$$x = \frac{13}{19}$$

Simplify any fraction if you can. Sometimes, you'll end up with a whole number. In this case, we can't simplify any further, so we've solved our problem.

Believe me, once you start doing these things and getting some practice, it becomes second nature. And that's what you want, because, well, who wants to spend too much time on this stuff?

> **POSITIVELY OOPS!**
>
> One very common mistake that occurs when solving equations is to incorrectly handle positive and negative signs when transposing. When the wrong sign is attached to a number, your answer will be "negatively" affected! So look out. It's important to distribute the correct signs across expressions, and then it's important to correctly change the signs whenever transposing a number or variable from one side of an equation to the other.

# YOUR TURN

Solve the following equations:

1. $(3x - 1) + 2 = (4x + 2) + 1$

2. $-7 - 3c = -5 - 2c$

3. $-5(1 + z) + 12z = 16$

4. $-3(6 - w) + 7w = 12$

5. $-9 - (-3 - 5t) = -3(7t + 4)$

6. $14x + 6 - [(3 - x) - (7x + 9)] = 15$

7. $19 - \{[3x - (x - 1)] - 5x\} = 0$

8. $4[3 - (2x + 4x) + 3x] = 6 + 2x$

9. $(4y + 7) + (3 - 2y) = 18y$

10. $3(4x - 2) + (3 - x) = 7$

# 5

# CHAPTER 5
## EQUATIONS WITH FRACTIONS

I'm sitting in the cafeteria thinking about solving equations with fractions, when I think, "I have to get a life." A life apart from algebra. Sometimes just hearing the word *algebra* gives you a little bit of a sweaty-palm feeling, right? Well, I'm here to tell you, "Nah, it's okay."

But I still have to get a life. Which is not easy to do when you can't form new memories. But I'm going to do it. And I'm going to do it by asking out the new girl, Liesa. She's French—a foreign exchange student. Being new, not only to the school, but to America, she's almost on the same footing as me. I have no memory, so I don't really have a past either. That's one reason to ask her out. Another is that she's gorgeous—just look at her sitting there in her cheerleader's uniform. I sort of have a thing for cheerleaders. . . . Yes, get a life, I know.

Do it, Ryan. Do it now. Go over there and say hi to her before you forget. Wait. She's getting up. She's taking her tray to the garbage. Do it now!

"Ryan."

No, what is it? What was I doing?

"Yo, Ryan." It's the cafeteria worker. "Are you going to take this tray you ordered, or what?" He extends a tray of coleslaw and beef Stroganoff invitingly.

"Um, I don't know. I actually feel kind of full. . . ."

"Listen, boy. Are you making fun of me?"

"No," I say. "Of course not. I have this problem where I can't form—"

"Yeah, I know. New memories. You told me. Look, just take your lunch and get outta here."

I take my tray and look around the room: a sea of unfamiliar faces. Ahh, but I am familiar with one thing: algebra—and don't laugh. Look, just because you don't like math doesn't mean you get to skip it for the rest of your life, you know. But it's okay. I'm here to help.

# FRACTIONS

To solve equations with fractions, you follow the same rules you learned in Chapter 4:

1. Simplify.
2. Transpose.
3. Simplify.
4. Divide by the coefficient.

Here's an example of a variable in the numerator of the fraction (that's the top number). Follow along to see how it's done.

$$4(x+3) - 2x = \frac{x+1}{2} + 10$$

Remove parentheses. $$4x + 12 - 2x = \frac{x+1}{2} + 10$$

Next, you want to get rid of the fraction. In order to do that, you have to multiply the entire equation (yes, both sides) by 2 so that the 2 in the denominator can be canceled out and your fraction will be gone. Here's how it works:

| | |
|---|---|
| Multiply both sides by 2. | $2\left(4x + 12 - 2x = \frac{x+1}{2} + 10\right)$ |
| This cancels the denominator. | $8x + 24 - 4x = \frac{\cancel{2}(x+1)}{\cancel{2}} + 20$ |
| And gets rid of your fraction. | $8x + 24 - 4x = x + 1 + 20$ |
| Group the like terms. | $4x + 24 = x + 21$ |
| Transpose. | $3x = -3$ |
| Divide by the coefficient. | $\frac{\cancel{3}x}{\cancel{3}} = -\frac{3}{3}$ |
| Change to a proper fraction. | $x = -1$ |

FRACTIONS

# VARIABLE IN THE DENOMINATOR

Now you are ready to solve an equation with fractions that have the variable in the denominator (that's the bottom number). This type of equation is called a **fractional equation.**

Don't worry: you follow the same four golden steps. It's just that this time, you take your answer for *x* and check it by placing it in the original equation. Here's our practice problem:

$$\frac{3+x}{x} = 5$$

Simplify (clear fractions).

$$x\left(\frac{3+x}{x} = 5\right)$$

$$x\frac{3+x}{x} = 5x$$

$$\cancel{x}\frac{3+x}{\cancel{x}} = 5x$$

$$3+x = 5x$$

Transpose. $3 = 5x - x$

Simplify. $3 = 4x$

Divide by coefficient.

$$\frac{3}{4} = \frac{\cancel{4}x}{\cancel{4}}$$

$$\frac{3}{4} = x$$

We've got an answer. Now, let's put it back into the equation in place of *x* and see if it works. It will probably be easier to work with the fraction as a decimal, so let's put .75 in place of *x*.

$\frac{3.75}{.75} = 5$ Simplify the improper fraction.

$5 = 5$ Our answer checks out. Good.

FRACTIONS

## TRY AGAIN

Let's try another problem that has two fractions, both with a variable in the denominator:

$$\frac{3}{x} + 3 = \frac{-1 - x}{2x}$$

Simplify by multiplying the entire expression by the common denominator of the two fractions, 2*x*:

$$2x\left(\frac{3}{x} + 3 = \frac{-1 - x}{2x}\right)$$

$$2\cancel{x}\left(\frac{3}{\cancel{x}} + 3\right) = \cancel{2x}\left(\frac{-1 - x}{\cancel{2x}}\right)$$

All fractions are clear. $6 + 6x = -1 - x$

Simplify. $6x + x = -1 - 6$

$$7x = -7$$

$$x = \frac{-7}{7}$$

$$x = -1$$

Now, put the answer -1 back into the original problem in place of *x* to check your work.

$$\frac{3}{-1} + 3 = \frac{-1 - (-1)}{2(-1)}$$

$$-3 + 3 = \frac{-1 + 1}{-2}$$

$$0 = \frac{0}{2}$$

$$0 = 0$$

Any 0 in a fraction makes the fraction equal to 0, so our answer checks out.

# CROSS MULTIPLY

Now's about the time when you're wishing for a shortcut, right? Well, there is something that can help. If you're doing a fractional equation with two fractions, one on each side of the equal sign, there is a shortcut you can use. It's called cross multiplying. Here's the kind of problem you need in order to do it:

$$\frac{5}{x-1} = \frac{2}{x}$$

Just multiply across the fractions, in the shape of an *x*: the numerator of the first fraction times the denominator of the second fraction. On the other side of your equal sign, you multiply the denominator of the first fraction by the numerator of the second. Like this:

$$\frac{5}{x-1} \times \frac{2}{x}$$

$$5x = 2(x-1)$$
$$5x = 2x - 2$$
$$5x - 2x = -2$$
$$3x = -2$$

$$x = -\frac{2}{3}$$

### COMMON MISTAKES

I've seen it happen countless times. You have a fractional equation staring you right in the face. You decide to get rid of that fraction by multiplying the whole equation by that pesky number in the denominator. That should clear out your fraction so you can go about solving it with ease, right? But do you know how many people only multiply across one side of the equation? That sets the whole problem off, and you've forever got an unbalanced equation after that. So, don't forget to multiply the *whole* equation by that number to clear the fraction.

# YOUR TURN

Solve the following equations:

1. $x - \frac{x+1}{2} = 4x - (1 + 2x) - \frac{x}{2}$

2. $10 - \frac{3x}{5} = x - 12$

3. $2(x - 2) = 5 + \frac{x-1}{4}$

4. $\frac{5}{x} = \frac{1}{2} + \frac{2+x}{x}$

5. Check your answer to problem 4. Replace $x$ with your answer.

6. $13 + \frac{3}{x} = 12 - \frac{4}{x}$

7. Check your answer to problem 6. Replace $x$ with your answer.

Cross multiply:

8. $\frac{8+3}{x} = \frac{5-2}{2}$

9. $\frac{15}{8x} = \frac{2}{9+x}$

10. $\frac{26}{8-r} = \frac{2}{4+r}$

# 6

# CHAPTER 6
## EQUATIONS WITH TWO VARIABLES

Look, forget about algebra just for a moment. I promise you, I can help. Yes, I'm Ryan, and yes, I'm a tutor, and yes, you've come to the right place. You need to get a good grade, I know.

But here's what's on my mind right now. Eiffel Tower. Baguette. Black coffee. Poodles. Give up? I'm going to ask out the French girl. She's here in the cafeteria with us—the one in the cheerleader's uniform, over at that table. This is a big deal for me. Since my accident, I haven't had much of a social life, you know what I mean? It's not easy when you *can't form new memories*!

Enough is enough. I'm going over there right now to strike up a conversation with her.

"Ryan."

What? What's this?

"Ryan, you forgot to get your lunch." It's the cafeteria worker.

"Oh, sorry. I have this condition where I can't form new memories."

"Indeed. Well, here's your tray of beef Stroganoff and coleslaw."

"Thank you." I take it back to my table. But there's no place to put my tray, because there's already an identical tray of beef Stroganoff and coleslaw sitting in my place, untouched. I throw a questioning glance at the cafeteria worker.

He shrugs sheepishly.

"I thought you said I didn't get my lunch?"

"Well, look, man, they're both yours."

"Why would you give me two lunches?"

"When the manager found out about your condition, he said to keep trying to sell you extra beef Stroganoff and coleslaw."

"Huh?"

"We made too much, it's going bad, whatever."

"I appreciate your honesty."

"You're not going to remember, anyway."

"You don't have to be that honest."

But I'll be honest . . . algebra is a lot easier to deal with than beef Stroganoff. By the way, if there's some section of this book that you need to go to for help, you don't have to wait for me. You can go right to it. Whatever works for you is fine.

# X AND Y IN THE SAME BOAT

Now we're getting into problems that "look" really tough—algebra equations with two different variables, such as an *x* and *y*.

$$4x + 2y = 3x - 2y$$

Isolate all the *x* and *y* variables on each side of the equation. Follow the same rules you have been using all along: simplify, transpose, simplify, and divide by the coefficient. Once you have the variables separated, you can solve for one of them. If you solve for *x*, the *y* will be in your answer. If you solve for *y*, the *x* will be in your answer. Check it out:

$$4x - 3x = -2y - 2y$$

$$x = -4y$$

And if you solve for *y*, the *x* will be in your answer:

$$4x - 3x = -2y - 2y$$

$$x = -4y$$

$$\frac{1}{4}x = y$$

# FINDING X AND Y

Okay, suppose you want to find the answer for both *x* and *y*. Let's look at the example:

$$x + y = 6$$

Well, if you think about it, there are several possibilities here. Perhaps $x = 2$ and $y = 4$. But what about $x = 5$ and $y = 1$? And $x = 3{,}000$ and $y = -2{,}994$. And don't forget $x = 5.25$ and $y = .75$.

Now, that's just insane. There are an infinite number of solutions for that problem. The same would be true for any equation with two variables. We're not interested in finding a ton of different solutions to the same problem. We'd really like to find just one answer that works. This is called a **linear equation in two variables**, and it looks something like this:

$$ax + by = c$$

With this type of problem, both *a* and *b* cannot equal 0. But ya know what? One of them can equal 0.

# SELECTING A VALUE FOR X

Let's look at this equation:

$$2x - y = 6$$

How can we find some solutions to this equation? Let's just select a value for one of the variables. How about *x*? You have to start somewhere, right? When we choose a value for *x*, we can plug it into the equation and find the value for *y*.

Let's plug in a 3 in place of *x* and see what happens:

$$2x - y + 6$$
$$2(3) - y + 6$$
$$6 - y = 6$$
$$y = 0$$

This equation works if $x = 3$ and $y = 0$. Now, try it again with *x* as a 5:

$$2x - y = 6$$
$$2(5) - y = 6$$
$$10 - y = 6$$
$$y = 4$$

So, another solution to the problem is $x = 5$ and $y = 4$. It's easy to go back and check it:

$$2x - y = 6$$
$$2(5) - (4) = 6$$
$$10 - 4 = 6$$
$$6 = 6$$

# ORDERED PAIRS

There's a little shortcut for showing your answers to linear equations with two unknown variables: it's called an **ordered pair**. In an ordered pair, the values of the variables are usually written within a pair of parentheses, in alphabetical order of the variables. So, if $a = 7$ and $b = 6$, the ordered pair is (7, 6). If $z = 12$ and $t = 21$, the ordered pair is (21, 12).

Using the example $2x - y = 6$ from the previous section, for our first solution, we can write the ordered pair (3, 0). For the second solution, we can write the ordered pair (5, 4).

Remember that the order of the pair is important. The solution would not work if you reversed the order and substituted 0 for *x* and 3 for *y*.

## IF X, THEN . . .

When you solve equations with two variables, you're solving a **conditional equation**. That's an equation that's only true for certain values. The answer for *y* depends on the answer for *x*. The two variables are related to each other within the problem.

When you take the SAT or other exams, you'll find lots of these conditional problems. You'll see them when you answer logic questions. They're expressed as "if this happens, then that happens." It's the same principle as "if *x* equals this, then *y* equals that." Get used to this type of problem. You're going to see it a lot!

# YOUR TURN

1. Solve for $x$ so that $y$ is in your answer:
   $13x + 4y = 5x - 3y$

2. Solve for $n$ if $r = 0$:
   $6r + 4n = 12$

Find the missing number so that each of the ordered pairs below is a solution to this equation: $3x + 2y = 9$

3. (3, ?)

4. (–3, ?)

5. (?, 0)

6. (?, 12)

For each equation below, determine if the given ordered pair is a solution to the equation:

7. (3, 2) $2x + 3y = 12$

8. (10, 3) $y = 3x + 1$

9. (16, 4) $2a - 3b = 20$

10. (3, 5) $8g - 3h = 8$

7

# CHAPTER 7
## EXPONENTS



Listen, it's great that you could meet me here in the cafeteria. I don't always do so well at keeping appointments. I have this condition: I can't form new memories. But I can remember all my algebra classes. Sometimes people freak out when they see the word *algebra*. Or when they see the letter *x*. I'm here to say, "Nah, it's not a problem."

No problem at all. Hey, do you mind waiting for a moment while I go talk to the new girl? The French student? Why? Uh, somebody told me she was looking for an algebra tutor. Got to keep up the business. Just because you guys don't like math doesn't mean you get to skip it for the rest of your life. Here I go.

Hold on. Why are there seventeen identical trays of beef Stroganoff and coleslaw at this table? No: ignore.

"Ryan."

"Shut up. . . . Um, excuse me. Liesa?"

The pretty girl in the cheerleader's uniform looks up at me and smiles. Her eyebrows raise, questioning.

"It is Liesa, right? I'm not that good with names. *Believe me*."

Still that puzzled look. But she's smiling. That's good. Um, now she's laughing. I look around. The kids at the other tables are all staring at me.

"*Ca va*, Ryan?"

*"Qu'est-ce qui se passe?"*

*"Bonjour, Ryan, ça va? Qu'est-ce qui se passe?"*

*"Arrete! Parle avec moi!"*

*"Tu dois lire ce livre."*

*"T'as oublié la stroganoff!"*

*"Aimes-tu le déjeuner?"*

"Yo, Ryan, are you going to pick up this coleslaw, or what, man?"

You know how quickly things can get confusing when there are numbers involved? Don't freak out. It's not a problem. Nah, it's really not a problem at all.

# EXPONENT BASICS

It will help to go over exponents so you're not just thrown right into algebra problems with exponents and square roots.

Here's an exponent:

$$5^2$$

The 5 is called the **base**, and the 2 is called the **exponent**. When you see a number with an exponent, it means you have to multiply the base by itself the number of times shown in the exponent. $5^2$ looks like this:

$$5 \times 5 = 25$$

Same goes for $3^4$. It doesn't represent $3 \times 4$. In fact, it is a short way of writing:

$$3 \times 3 \times 3 \times 3 = 81$$

Any number to the 0 power equals 1. So, $8^0 = 1$. Any number to the first power equals that number. So, $8^1 = 8$.

# OPERATIONS WITH EXPONENTS

In order to add or subtract numbers with exponents, solve for each exponent first and then continue with your problem:

$$5^2 + 5^4 = (5 \times 5) + (5 \times 5 \times 5 \times 5) = 25 + 625 = 650$$

When you multiply exponents, check to see if the bases are the same. If they are, just add up the exponents and keep the base the same.

$$3^2 \times 3^4 = 3^6$$
$$3 \times 3 \times 3 \times 3 \times 3 \times 3 = 729$$

When the bases are different, solve for each exponent individually and then go on with your problem:

$$3^2 \times 4^4 = (3 \times 3) \times (4 \times 4 \times 4 \times 4) = 9 \times 256 = 2{,}304$$

The same thing goes for division problems with exponents. If the bases are the same, you can subtract the exponents and keep the base the same. If the bases are different, simplify the exponents first, and then do your division problem.

$$5^3 \div 3^2 = (5 \times 5 \times 5) \div (3 \times 3) = 125 \div 9 = 13.89$$

# ORDER OF OPERATIONS

What happens when your equation or expression has more than just exponents in it? What do you do first? Just think back to the order of operations: Please Excuse My Dear Aunt Sally. That means you should do your equation in this order: Parentheses, Exponents, Multiplication, Addition, and Subtraction.

So, after you do any operations inside the parentheses (including solving exponents), you should go through and solve for any other exponents in the problem. Here's a sample:

$$2(3x + 2^2) = 4x - 3^2$$
$$2(3x + 4) = 4x - 9$$
$$6x + 8 = 4x - 9$$
$$6x - 4x = -9 - 8$$
$$2x = -17$$

$$x = -\frac{17}{2}$$

$$x = -8.5$$

It's best to check your work to see that you did it right. Put -8.5 in place of the $x$ throughout the problem:

$$\begin{aligned} 2(3x + 2^2) &= 4x - 3^2 \\ 2(3 \cdot -8.5 + 2^2) &= 4 \cdot -8.5 - 3^2 \\ 2(-25.5 + 4) &= -34 - 9 \\ -51 + 8 &= -43 \\ -43 &= -43 \end{aligned}$$



# ROOTS

Remember what a root, or a square root, is? It's kind of the opposite of an exponent. A **root** is a divisor of a quantity that gives you that quantity when it's squared. Since $4^2 = 16$, the square root of 16 is 4. Roots are written like this:

$$\sqrt{16} = 4$$

The square root of 16 is a perfect square. The sign around the 16 is called the **radical sign**.

Very often, you'll see a number, or coefficient, in front of a square root sign. This coefficient needs to stick to the radical and should be multiplied by any square root once it's solved:

$$3\sqrt{4} = 3(2) = 6$$

Since we can tell that the square root of 4 is 2 (because $2^2 = 4$), we multiply the coefficient of 3 times the perfect square of 2, and we get the answer 6.

# SIMPLIFYING ROOTS

You might wonder what happens when your root does not work out to be a perfect square. Well, we can't get rid of it, but we can simplify it. Ask yourself what numbers can be factored out to get to perfect squares. So, when you are looking at the square root of 40, you're also looking at the square root of 4 times the square root of 10. Watch:

$$\sqrt{40} = \sqrt{4} \cdot \sqrt{10}$$

Remember, it's not the same as saying that the square root of 40 equals 4 times 10. You have to keep the radical sign over the numbers until you can find a perfect square.

Since $\sqrt{4}$ is 2, you can "pull out the square root" from the equation and have the $\sqrt{4}$ be represented by a 2. That way, you can represent it as a coefficient, like this:

$$\sqrt{40} = 2\sqrt{10}$$

Notice that there were other possibilities to choose from to simplify $\sqrt{40}$. You could have chosen $\sqrt{5} \cdot \sqrt{8}$. But that doesn't leave you with any perfect squares. The idea behind simplifying roots is get as many numbers as possible out from under the radical signs. So, think before you choose which numbers you use to simplify a root.

EXPONENTS

### GROWING, GROWING, GROWN

When you increase a number by an exponent and then increase that number by an exponent as well, that number will grow out of control faster than a deadly runaway virus. Say you donate 2 cents one day to a local charity. Then you increase the amount exponentially each day—that is your donation is squared, or raised to the second power. In just a few months, you'll be donating more than a trillion dollars a day. Don't get yourself into that kind of debt! In fact, if you ever see a bank offer interest in such a manner, grab it!

# YOUR TURN

1. Rewrite this expression with exponents, then solve it:
   $(3 \times 3 \times 3 \times 3) + (2 \times 2 \times 2)$

Solve these problems. Use the order of operations when necessary:

2. $5^5 + 8^4$

3. $4(2x + 2^2) + 4x = 6x - 3^2$

4. $(8 + 3^3) + (4 - 5^4) = x + 20$

5. $x(3^2 + 2^3) \cdot 3x(5^2) = 50$

Simplify these expressions:

6. $\sqrt{81} + \sqrt{169}$

7. $\sqrt{147}$

8. $5\sqrt{12} - 13\sqrt{27} + 17\sqrt{45}$

9. $\sqrt{5}(2 - \sqrt{3})$

10. $-2\sqrt{12} \times (-5\sqrt{3})$

# 8

# CHAPTER 8
## THE LAWS OF EXPONENTS

I'm glad you've come to me for help. Your report card will thank you. I know, you probably freak out a bit when you see the word *algebra* or when you see the letter *x*. I'm here to tell you. . . .

What? You think I should ask her out? The new girl? But she just started today. Besides, she's French. Not to mention a cheerleader. That's a little out of my league, don't you think?

Okay, fine. You don't have to tell me to get a life. It's a little unkind. I have a disability, after all. Oh, you think it's time I stopped hiding behind my disability? Time that I joined the human race?

Maybe you're right. Let me just pick up this tray of coleslaw—no? You'll get it for me? How thoughtful.

"Liesa? Your name is Liesa, right? I'm a little bad with names. I have this condition—"

"Yes, you freak, my name is Liesa. Sit down. Put that tray anywhere, okay? No, I'm not French, by the way. Let's just drop that little game for two minutes, okay, sweetie?"

"Um . . ."

"You want to ask me something, Ryan? You want to ask if I'll go out with you? You want to date me—is that it?"

"Well . . ."

"Spit it out. How are you going to get a life if you don't say what you want?"

"Well, I—"

"Look, let me ask you a question. If I agreed to go out with you, how would you remember to meet me?"

"Well—"

"No, here's a better question. How do you know we haven't dated *already*?"

"Uh, I—"

"How do you know I'm not already *your girlfriend*? Look, I'm going to go take this tray over to the garbage. I want you to sit here and think about this for a minute and let me know when you've got your answer straightened out."

Well, that went well, I think. Look, just because you don't like math doesn't mean you can skip it for the rest of your life. Yes, get a life, I know.

# EXPONENT LAWS

You are about to enter into the next exciting phase of algebra. You've mastered expressions and equations. You will soon be a master of exponents—and with that knowledge, you can plow through just about any algebra problem that's thrown your way. There are five laws that you need to learn and practice before you become proficient with the "power."



# MULTIPLICATION LAW

$$b^n \cdot b^m = b^{n+m}$$

I know you just learned this law in the last chapter, but let's go over it again because there are some funky combinations that we didn't talk about that might be worth thinking about now. The setup of numbers will not always look as straightforward as it does in the above example.

Say, for instance, you have a number to the first power. There won't be an exponent showing above the number:

$$3 \cdot 3^4$$

So, you have to recognize that multiplying these numbers together means multiplying two exponents. Treat the 3 as $3^1$ . Then follow the multiplication law:

$$3 \cdot 3^4 = 3^1 \cdot 3^4 = 3^{1+4} = 3^5$$

Another example is parentheses and exponents. Remember that parentheses mean multiplication. So,

$$b^3(4b^2) = b^3 \cdot 4 \cdot b^2 = 4b^5$$

Oh, and here's a good one that you might not even recognize as something that would fit under the multiplication law:

$$(x^4y^2)(xy^3)(6xy)$$

Remember, the multiplication law applies for expressions with more than just two products. Separate each variable individually so that you can regroup them and add them up.

$$x^4 \cdot y^2 \cdot x^1 \cdot y^3 \cdot 6 \cdot x^1 \cdot y^1$$
$$6x^{4+1+1} \cdot y^{2+3+1}$$
$$6x^6y^6$$

# POWER OF A POWER LAW

$$(b^n)^m = b^{nm}$$

This one is pretty easy, and it's a nice shortcut too. When you see a power raised to another power, you multiply the exponents together.

$$(7^2)^3 = 7^{2\times 3} = 7^6$$

The parentheses should be an easy reminder to multiply.

If we try to solve this expression the long way, we have to use the multiplication law to get the same solution:

$$(7^2)^3 = (7^2)(7^2)(7^2)$$
$$7^{2+2+2} = 7^6$$

# POWER OF A PRODUCT LAW

$$(bc)^n = b^n c^n$$

This law is almost exactly like the power of a power law, but it's really a reminder that if you have another variable or number inside your parentheses, it has to be changed to the same power that the variable is being changed to:

$$(4r^2)^2 = 4^2 r^4 = 16r^4$$

Even though the 4 is not expressed to any power other than 1, it has to be squared along with the *r*.

Sometimes, you'll find that both the terms inside the parentheses are expressed to a power:

$$(3^2 x^3 y^4)^2 = 3^4 x^6 y^8$$

In this case, each product in the expression is raised to the power that's shown on the outside of the parentheses.

# POWER OF A FRACTION LAW

$$\left(\frac{a}{b}\right)^n = \frac{a^n}{b^n}$$

If you have a fraction such as $3/4$ and it has to be raised to some power, you must raise both the numerator and the denominator of the fraction to that power, like this:

$$\left(\frac{3}{4}\right)^2 = \frac{3^2}{4^2} = \frac{9}{16}$$

Of course, once you get your answer, you should simplify the fraction if possible. But this one looks like it's already simplified as far as it can go.

This same rule applies to fractions even when the numerator and denominator are already raised to a power. You're really just trying to remove parentheses when you perform the power of a fraction law, and removing parentheses means multiplication. Check out this example. We use *x* and *y* in the fraction instead of real numbers, because the numbers would get huge if we really did the math.

$$\left(\frac{x^3}{y^2}\right)^3 = \frac{x^{3 \cdot 3}}{y^{2 \cdot 3}} = \frac{x^9}{y^6}$$

# DIVISION LAW

$$\frac{b^m}{b^n} = b^{m-n}$$

Need another example?

$$\frac{6^2}{6^3} = 6^{2-3} = 6^{-1} = \frac{1}{6}$$

And just for good measure, here's a third.

$$\frac{2^4}{2^2} = 2^{4-2} = 2^2 = 4$$

# MIXING IT UP

Now, why learn five different laws if you're not going to combine everything and mix it all up? That's the fun and the challenge of learning something new. You can put it to use in a problem like this:

$2^3(2 \cdot 4^2)^2$

$2^3(2^2 \cdot 4^4)$ Use the power of a product law.

$2^3 \cdot 2^2 \cdot 4^4$

$2^{3+2} \cdot 4^4$ Use the multiplication law.

$2^5 \cdot 4^4$

$32 \cdot 256$

$8{,}192$

Here's another one to follow through:

$(4^2)^2 \times \left(\frac{6^2}{6^3}\right)^2$

$4^{2 \cdot 2} \times \left(\frac{6^2}{6^3}\right)^2$ Use the power of a power law.

$4^{2 \cdot 2} \times \frac{6^{2 \cdot 2}}{6^{3 \cdot 2}}$ Use the power of a fraction law.

$4^4 \times \frac{6^4}{6^6}$ Simplify.

$4^4 \times \frac{\cancel{6^4}}{6^{\cancel{6}}{}_2}$ Cancel out.

$256 \times \frac{1}{36}$

$\frac{64}{9} = 7.11$



## PERFECT CUBES

When you think of a cube, you probably think of something with four sides, like a square or a box. Well, go figure, but when something is **cubed** in math, it's brought to the third power, not the fourth. It's very common to **cube** a number, so here's a handy list of numbers that are raised to the third power:

$1^3 = 1$ $6^3 = 216$

$2^3 = 8$ $7^3 = 343$

$3^3 = 27$ $8^3 = 512$

$4^3 = 64$ $9^3 = 729$

$5^3 = 125$ $10^3 = 1{,}000$

# YOUR TURN

Which law of exponents must be used to solve these problems?

1. $4 \cdot 4^5 = 4^6$

2. $(2^3)^2$

3. $(5 \cdot 7)^3 = 5^3 \cdot 7^3$

Simplify each expression:

4. $5x^0$

5. $x^2(x^4)^5$

6. $(4x^2)^3(2x^2)^2$

7. $(x^3y)(xy^2)$

8. $\left(\frac{2^2}{3^3}\right)^2$

9. $\frac{5^4}{5^2}$

10. $\frac{(2xy)^2}{(3x^2)^3}$

THE LAWS OF EXPONENTS

# 9

# CHAPTER 9
## MULTIPLYING POLYNOMIALS

Sorry, no time for algebra today. Tonight is *prom night*!

I bet you never thought a freak like me would get a date. I bet you thought a guy who *can't even form new memories* would probably have a bit of trouble lining up a date for the prom, right? You probably thought maybe one of my students, a sophomore with thick glasses who'd skipped two grades in elementary school and studies algebra to get ahead, would feel sorry for me and agree to go with me, right?

No, no. I'm going with a cheerleader. A *French* cheerleader. The very beautiful *new girl*. The girl everyone's been wanting to ask out. Yes, I know, they all had prom dates already when she arrived. Luck was on my side.

So I pick her up. Her mother and father are on the front lawn to take pictures of us in our formal attire as I pin on her corsage. I mean, the mother and father of her host family, of course. My heart is in my throat as I drive her to school. My palms are only the slightest bit sweaty as I walk through the parking lot toward the school gym. And am I fooling myself that her palms are just the slightest bit sweaty too? (It's hard to tell because my palms are sweaty.)

The door swings open; I see lights and hear the noise of a crowd inside. I slip her hand under my forearm and lead my queen triumphantly into. . . .

Wait. It's way too brightly lit to be the prom. The floor is filled with cheerleaders, not dancers. The bleachers are full of students, and they aren't in formal wear. Liesa is gone—gone to join the cheerleaders in the halftime show—of which I, it seems, am the star attraction. The entire school erupts in laughter.

You know, I'm glad I'm not going to remember this. I slip into a seat in the bleachers and hug my Gingiss tux jacket around myself. Sometimes people freak out. Like when they see the letter *x*. But I'm here to tell you. Not a problem. No problem at all.

# GEARING UP

Okay, you know how to deal with many situations in algebra, including exponents, fractions, groupings in parentheses, all sorts of stuff. So, rather than have you write out expressions, distribute numbers, and show every step every time, I'll show you a good shortcut for multiplying two binomials together. But, of course, before you learn the short version, you have to learn the long version.

# MULTIPLYING BINOMIALS

To multiply binomials you have to use the distributive property, but instead of distributing just one number or variable, you have to distribute a whole expression. Let's see how that works:

| | |
|---|---|
| $(x+2)(x+7)$ | |
| $(x+2)(x)+(x+2)(7)$ | Distribute $(x+2)$. |
| $x^2+2x+7x+14$ | Multiply. |
| $x^2+9x+14$ | Combine terms. |

Try it one more time to make sure you've got it.

| | |
|---|---|
| $(3x-2)(x+5)$ | |
| $(3x-2)(x)+(3x-2)(5)$ | Distribute $(3x-2)$. |
| $3x^2-2x+15x-10$ | Multiply. |
| $3x^2+13x-10$ | Combine terms. |

Got it? Good. Now learn a faster way to do it. It involves doing a little work in your head, but things will be faster and smoother once you get it.

# FOIL METHOD

Here's the shortcut for multiplying two binomials:

1. Find the product of the two **first** terms of the binomials.
2. Find the product of the two **outermost** terms.
3. Find the product of the two **innermost** terms.
4. Find the product of the two **last** terms.

We can use a good memorization aid to remember those steps. **F**irst, **O**uter, **I**nner, **L**ast, or **FOIL**.

So let's look at the first example again:

$$(x + 2)(x + 7)$$

This is what the terms look like:

first: $(x + 2)(x + 7)$ The first terms are $x \cdot x$ or $x^2$.
outer: $(x + 2)(x + 7)$ The outer terms are $7x$.
inner: $(x + 2)(x + 7)$ The inner terms are $2x$.
last: $(x + 2)(x + 7)$ The last terms are $2 \times 7$ or 14.

So, we can rewrite $(x + 2)(x + 7)$ as

$$(x + 2)(x + 7)$$
$$x^2 + 7x + 2x + 14$$
$$x^2 + 9x + 14$$

As you keep doing this and get more practice, you will see that you end up with a $2x$ and $7x$ and can skip a step and add them right away.

# SQUARING A BINOMIAL

Here's another rule to remember:

$$(a + b)^2 = a^2 + 2ab + b^2$$
$$(a - b)^2 = a^2 - 2ab + b^2$$

Let's try it with $(7y + 6)^2$:

| | |
|---|---|
| $(7y + 6)^2$ | Square the first and last terms. |
| $(7y)^2 + 2(42y) + 6^2$ | Double the product of the middle terms. |
| $49y^2 + 84y + 36$ | Simplify. |

# MULTIPLYING POLYNOMIALS

When you multiply two polynomials, you might need to go back to the distribution method. When you have a trinomial and a binomial, you have to distribute each term in the binomial so that it gets factored with each term in the polynomial.

| | |
|---|---|
| $(5x + 4)(2x^2 - 3x + 7)$ | |
| $10x^3 - 15x^2 + 35x + 8x^2 - 12x + 28$ | Distribute. |
| $10x^3 - 7x^2 + 23x + 28$ | Combine terms. |

Do this for as many terms as necessary. This particular problem distributes out to six terms, but then you can combine the like terms at the end.

## MEMORY TRICKS

A **mnemonic** is a system meant to improve our memory. It can be a formula, a rhyme, or an acronym like FOIL. It's a lot easier to memorize the word *foil* than to remember "first, outer, inner, last." You're much more likely to forget the words or to remember them in the wrong order if you don't use a mnemonic aid. Something like *foil* not only tells you the order in which you should do steps in an algebraic problem; it's also easier to remember because you can picture something in your head.

# YOUR TURN

Simplify the following expressions:

1. $(6x + 2)(x + 9)$

2. $(4x + 3)(x - 1)$

3. $(5b + 3)(3b - 2)$

4. $(6z - 4)(6z + 4)$

5. $(12xy - 2)(3xy + 7)$

6. $(5y + 7)^2$

7. $(8z - 5)^2$

8. $(x + 1)(x^2 - 3x + 2)$

9. $(3x^2 + 5x + 1)(2x - 1)$

10. $(y - 1)(3y^2 + 4y + 2)$

# 10

# CHAPTER 10
## WORD PROBLEMS

What do people really mean by "get a life"? What, exactly, does a life consist of? Some would say friends. Some would say *girl*friends. But what about family? What about a mother's love?

What about a sense of purpose? Having something that you're good at—like algebra. Being able to help others, like with tutoring.

Here I am, in the school hallway. Life moves all around me, a river of bodies. Sometimes I think all I have to do to get a life is reach out and touch it. How beautiful life is! Just look at that girl in the cheerleader's uniform.

"Put your hand down, Ryan. Do you think I'd let you touch me? You're a freak. A *freak,* do you hear me? How could you? How could you *do it*? You're disgusting. You. Disgust. Me. And you did it for what? For *her*. Your taste in women is pretty sick, Ryan. That's all I'm going to say. No, it's not. I have something else to say. You *humiliated* me, Ryan. And now I'm going to humiliate you. I'll tell you exactly what I'm going to do to you, and you're not even going to remember—you freak! For starters, I'm putting on a one-man freak show at the big game tomorrow night, and you're the star attraction. You're going to Gingiss to rent a tux tonight. Don't worry—I put the order form and 'prom' invitation in your locker so you won't forget. I know how you like to get things in writing. (Freak.) Then I'm going to tell all my friends in the French club just what a *freak* you are. We'll have a little something in store for you."

I know what Shakespeare has to say about getting a life: "And yet to me, what is this quintessence of dust? Man delights not me—no, nor woman neither." And just because you don't like math doesn't mean you get to skip it for the rest of your life, okay?

# PROBLEMS WITH A VARIABLE

Let's think for a minute about why word problems even exist in this world. They're meant to solve a problem or an unknown variable, just like algebra does. So, solving an algebraic word problem seems like the ultimate mastery of algebra, doesn't it?

Well, here's an example of what I mean:

Six times a number is decreased by 8, and the result is 4 more than 3 times the number. What is that number?

Let's rewrite that with $x$ plugged in as *a number*. We can also substitute words like *decrease* with a minus sign and *result* with an equal sign. You know: translate it into algebra.

$$6x - 8 = 4 + 3x$$

Now, let's solve the equation:

| | |
|---|---|
| $6x - 8 = 4 + 3x$ | |
| $6x - 3x = 4 + 8$ | Transpose. |
| $3x = 12$ | Combine terms. |
| $x = \frac{12}{3}$ | Isolate the variable. |
| $x = 4$ | Simplify. |

Don't you think it's a good idea to plug the 4 into the original problem in place of the $x$ to check our handiwork? It definitely is a good idea.

Our answer is correct: $x = 4$ because the problem works only when we plug 4 into the equation in place of $x$. The two sides of the equation balance, or equal each other.

$$\begin{aligned} 6x - 8 &= 4 + 3x \\ 6(4) - 8 &= 4 + 3(4) \\ 24 - 8 &= 4 + 12 \\ 16 &= 16 \end{aligned}$$

# WORD PROBLEM CHEAT SHEET

Here's a list of common word-problem phrases that translate into algebra. You'll see a lot of these over and over again.

| Word Problem Lingo | Math Translation |
|---|---|
| $x$ is equal to 5<br>$x$ is the same as 5<br>$x$ is 5<br>$x$ and 5 are equivalent | $x = 5$ |
| $x$ and 4<br>The sum of $x$ and 4<br>$x$ added to 4<br>$x$ increased by 4<br>$x$ more than 4 | $x + 4$ |
| Less than $x$<br>$x$ minus 7<br>$x$ take away 7<br>$x$ decreased by 7<br>$x$ less 7<br>The difference of $x$ and 7 | $x - 7$ |
| $x$ times 9<br>The product of $x$ and 9<br>$x$ multiplied by 9 | $9x$ |
| $x$ divided by 6<br>The quotient of $x$ and 6<br>6 divided into $x$ | $\frac{x}{6}$ |
| $\frac{2}{3}$ of $x$<br>27% of $x$ | $\frac{2}{3}x$<br>$.27x$ |

# WORD PROBLEMS WITH LINEAR EQUATIONS

Okay, let's try another word problem.

The sum of three consecutive integers is –102. What are the integers?

Should we set this up as a problem with three variables? No. The reason is that the problem asks for three consecutive numbers. That can be written as $x$, $x + 1$, and $x + 2$. That way, you only have to find $x$, and then you can find the consecutive numbers when you're finished.

$$x + (x + 1) + (x + 2) = -102$$
$$x + x + 1 + x + 2 = -102$$
$$3x + 3 = -102$$
$$3x = -102 - 3$$
$$3x = -105$$
$$x = -\frac{105}{3}$$
$$x = -35$$

So $x = -35$. But don't forget that you have to find three consecutive integers. Remember that adding negative numbers is a little like subtracting them. So, instead of the consecutive numbers being 35, 36, and 37, we have to stop and look at the signs first. Consecutive negative numbers would be –35, –34, and –33. You can plug them into the original equation to check your answers:

$$-35 + (-34) + (-33) = -102$$
$$-35 - 34 - 33 = -102$$
$$-102 = -102$$

# WORD PROBLEMS WITH TWO UNKNOWNS

Sometimes, word problems need to be read a hundred times before you can figure out what they're saying. But once you crack the code and figure out what they're really asking you, it's no problem writing out your equation and solving it like a pro.

The sum of two numbers is 36. Their difference is 6. What are the two numbers?

Whenever there's a relationship between the two variables, you can use an equation to express that relationship. For example, we see in the word problem that the difference between the two variables is 6. That's a real number that can be used to subtract one from another, so one of the variables can be expressed as $x - 6$. We also know from reading the word problem that the sum of the two variables is 36. Let's write the equation out and solve it:

$$x + (x - 6) = 36$$
$$2x - 6 = 36$$
$$2x = 36 + 6$$
$$2x = 42$$
$$x = \frac{42}{2}$$
$$x = 21$$

So, one number equals 21. If the other variable equals $x - 6$, then it should equal $21 - 6$, or 15. Let's plug those numbers into the original problem:

$$21 + 15 = 36$$

## PUZZLING PROBLEMS

Word problems are among the most feared of all math problems. This might be because word problems take a lot of concentration, and you are figuring out the answers to some rather odd and confusing questions. Lewis Carroll's classic book *Through the Looking Glass* expressed this type of frustration with nonsensical problems:

"Can you do addition?" the White Queen asked. "What's one and one and one and one and one and one and one and one and one and one?" "I don't know," said Alice. "I lost count."

Unlike Alice, you now know how to tackle word problems, so you won't lose count.

# YOUR TURN

Write expressions that show the following problems in mathematical terms:

1. One number is $\frac{4}{5}$ of another.
2. Margo is three times older than Sara.
3. Jeff has eight times as many quarters as Mike.
4. Two numbers are consecutive integers.
5. The width of a rectangle is seven more than three times its length.
6. Five times a number decreased by nine is equal to another number.

Solve the following word problems:

7. The sum of two consecutive integers is 39. Find the integers.
8. Nine less than three times a number is equal to three more than twice the number. Find the number.
9. Laura is six times older than Kate. The sum of their ages is 28. How old are Laura and Kate?
10. Find three consecutive integers such that the sum of the first two decreased by 11 is equal to the third.

# 11

# CHAPTER 11
## GRAPHING COORDINATES

"Ryan, listen to me. What I have to tell you is *trés* important."

I'm sitting in, of all places, Craig's living room. A room I haven't seen since the sixth grade. I'm sitting in one of those plush, antique chairs, across from Craig's mother. I'm holding a Wedgwood demitasse filled with coffee.

Craig's lovely mother. She's wearing a housedress with three-quarter-length sleeves turned up and buttoned at the cuffs, the kind women used to wear in France a hundred years ago. The dress is made of white silk damask, with a pattern of interlocking peacocks and birds-of-paradise. A white pair of Manolo Blahniks twine up her ankles like twin albino garter snakes. She shifts in her chair and puts down her coffee.

"Ryan, you have to be careful. Some people will try to take advantage of your disability. A girl like Liesa is no good for you. Stay away from her."

She pauses, regarding me thoughtfully. I can smell her perfume, quietly enveloping me in its lazily provocative way. *Arpège* by Lanvin, I think.

"I'm sorry about yesterday. You must think I'm crazy—or you would if you remembered. I wore that silly getup because I thought if I made you relive that day all those years ago, it might shock you into getting your memory back. Obviously, it didn't work."

I sip my coffee. The cup clatters nervously against the saucer.

"You're like a son to me, Ryan. You know that, don't you? I will never let anything serious happen to you. Even the principal is not your friend, but you don't need to worry about him. I have friends on the school board." She rises. "Take care of yourself, Ryan."

Then she leans over me. I smell coriander, heart of rose, mimosa, jasmine, and geranium. It makes me slightly dizzy.

She plants a kiss on my cheek.

Ryan Hall. Algebra tutor *extraordinaire*. I like the sound of that.

# GRAPH BASICS

I always thought graphs were pretty fun. They're a good break from solving endless problems, and it's kind of a visual thing instead of a mathematical thing.

In algebra, graphs are broken into four sections called **quadrants**. They're always numbered the same way. The top right is numbered I (with a Roman numeral), the top left is numbered II, the bottom left is III, and the bottom right is IV. Here's how that looks:

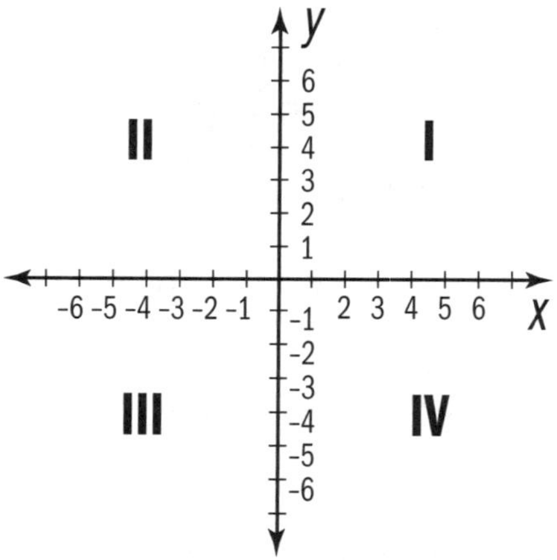


Also, the **horizontal axis**, or the line across, is usually called *x*. The **vertical axis**, or the line up and down, is usually called *y*. The point where they cross in the center is 0. That's because the *x*- and *y*-axes represent number lines. They show both positive and negative numbers, and 0 is where they both cross and meet up with each other.

Let's go over that:

1. The *x*-axis is a horizontal number line.
2. The *y*-axis is a vertical number line.
3. The center where the axes cross is 0.
4. The four quadrants are labeled I, II, III, and IV.

# ORDERED PAIRS

Remember back in Chapter 6, we talked about solving equations with two variables? In order to save space and time, the answers to each variable were written as an **ordered pair**. That looks something like this: (4, 7). It's written to represent $(x, y)$. That means in an equation with an $x$ variable and a $y$ variable, $x = 4$ and $y = 7$. So (4, 7) is faster and easier.

When you think of graphing, the whole ordered pair thing takes on new meaning. You can actually use the 4 and the 7 and plot them on a graph. Think about it. The 4 is a positive number. That means it belongs on the $x$-axis in the positive section. The 7 is positive and belongs along the vertical $y$-axis. If you plot these two numbers on the graph, you end up in quadrant I.

Here's how you plot the coordinates (4, 7):

1. Start at the 0 (the center, or origin).
2. Move along the $x$-axis 4 spaces.
3. Move up the $y$-axis 7 spaces.
4. Draw a dot. That's (4, 7) on the graph.

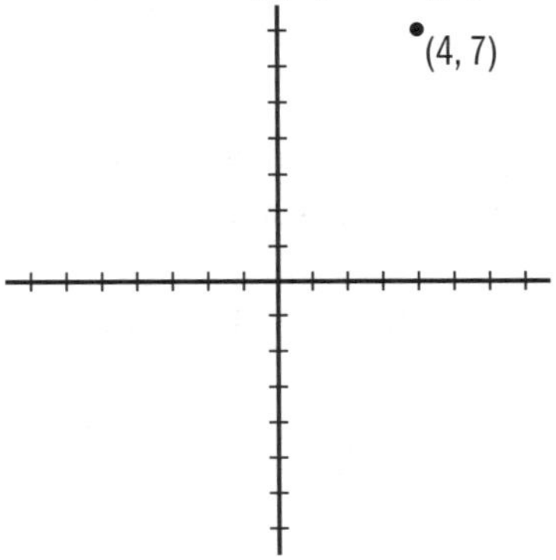

# NEGATIVE COORDINATES

When you have negative numbers in your ordered pair, that means you will end up with a point plotted in either quadrant III or IV. See what I mean. Plot (–3, –6).

1. Start at the 0 center, or origin.
2. Move along the *x*-axis negative 3 spaces.
3. Move down the *y*-axis negative 6 spaces.
4. Draw a dot. You're in quadrant III.

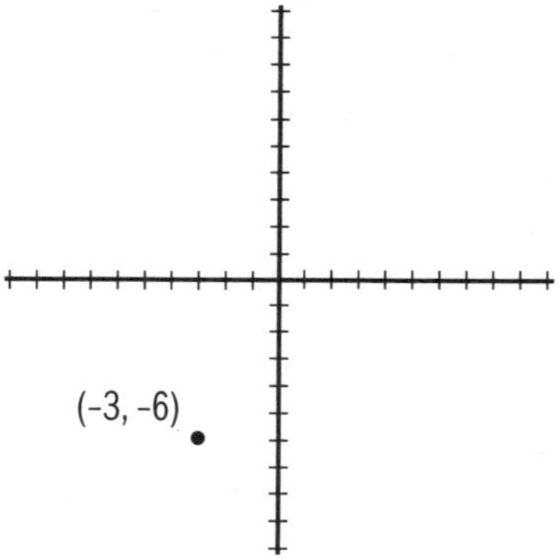


A good guideline to remember when plotting coordinates is to treat the *x*-axis and *y*-axis like number lines. On the *x*-axis, numbers to the right of 0 are positive. Numbers to the left of 0 are negative.

The same thing goes for the *y*-axis. Numbers above the 0 are positive. Numbers below the 0 are negative.

- If *x* is positive, move to the right.
- If *x* is negative, move to the left.
- If *y* is positive, move up.
- If *y* is negative, move down.

# LETTERED COORDINATES

Okay, you can probably imagine that this coordinate stuff gets pretty messy if you put several dots on the same graph. Here's the math-world solution to that problem: Letter the coordinate pairs.

For example, we want to plot the following coordinates:

*A* (-2, 0)
*B* (5, 2)
*C* (-1, -1)
*D* (6, -3)

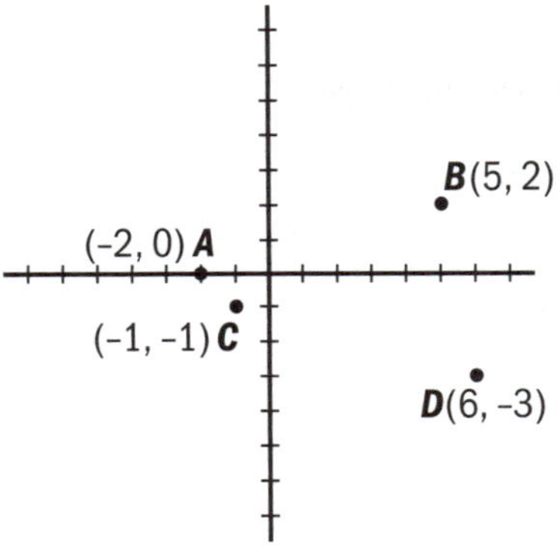


## CARTESIAN SYSTEM

Have you heard of the famous French philosopher and mathematician René Descartes? He really knew his stuff. We owe our whole graphing system to his work in the early half of the 1600s. He was the first person to use two perpendicular number lines (the *x*- and the *y*-axes) to plot points. In his honor, the graph with four quadrants that we use is named after him. It's called the **Cartesian coordinate system**.

# YOUR TURN

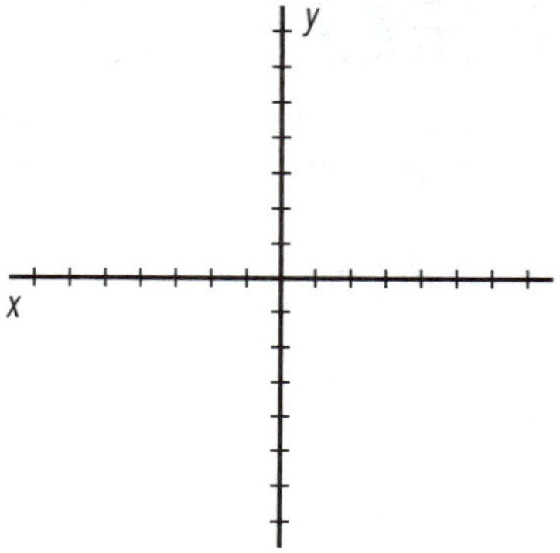


In which quadrant would the following ordered pairs be plotted?

1. (-2, 4)
2. (3, 5)
3. (2, -6)
4. (-3, -3)

Plot these ordered pairs:

5. A (3, 4)
6. B (-4, 2)
7. C (-5, -4)

What would the ordered pair be if you start at the center origin and do the following?

8. Move along the *x*-axis negative 5 spaces. Move along the *y*-axis positive 4 spaces.
9. Move along the *x*-axis 0 spaces. Move along the *y*-axis -3 spaces.
10. Move along the *x*-axis a positive half-space. Move along the *y*-axis negative 2 spaces.

# 12

# CHAPTER 12
## LINEAR EQUATIONS



Just walking up Craig's driveway is enough to send me soaring down memory lane. I'm hit by a hundred flashbacks of the two of us skateboarding there and nearly breaking our necks daring each other to do stupid tricks. I feel fine about tutoring Craig for the first time, but I have this nervous feeling in the pit of my stomach about seeing his mom again. I ring the bell and Craig answers.

"Look, Ryan, you can go ahead in and have your cookie or whatever, but I have better things to do."

"Now, Craig, just because you don't like math doesn't mean you get to skip it for the rest of your life. I know you probably want to be out playing basketball, but your report card will thank you."

"Dude, I'm in calculus now. Algebra was *four years ago*. I haven't been on the basketball team since junior high. The only reason you come here every day is that my mom feels sorry for you. Jeez, it's like talking to a brick wall. I'm outta here."

Craig's living room looks just the same as it did—plush antique chairs upholstered in silk, comfy sofas, flowers on the side tables.

"Oh, Ryan, did that idiot boyfriend of mine leave?"

I turn around to confront a very pretty girl in a cheerleader's uniform.

"Is it true you have no memory, or is all that just a game?"

"Do we know each other?"

"Sit down, Ryan. Yes, here on this nice sofa. No, next to me. Now, if you really have no memory, we can do whatever we want, right? No one will ever find out. No consequences." She twines her arms around my neck. "Come here . . ."

"*Liesa!* Ryan. I'm surprised at you!"

There, in the doorway, bearing a silver tray with two cookies and two half-full glasses of milk, stands Craig's mom. *Wearing a cheerleader's uniform.* With its River Dell High colors and bold *X* insignia, it's identical to that worn by the girl on the sofa next to me, whom I've never seen before in my life.

My palms are sweaty. My tongue feels thick. The three of us look at each other for a long moment. Finally, I break the silence.

"Er, Mrs. G. You look . . . just as beautiful as you did when we were kids."

Liesa sobs and runs out the front door.

Some people freak out when they see the letter *x*. I'm here to tell you, Nah, it's not a problem.



# LINEAR EQUATIONS

Now that you know the basics of graphing, you can solve equations, plot them on a graph, and make a straight line. Well, you can't just solve *any* old equation and plot it in a straight line. There are certain rules, as you have learned to expect in algebra. (There are always rules!)

In order to plot a line on a graph from an equation, you have to solve a linear equation. Here's what's involved in a linear equation:

1. There are two variables.
2. The variables are involved with the four basic operations (addition, subtraction, multiplication, or division).
3. Neither variable is raised to a power (other than 1).
4. Neither variable is in a denominator of a fraction.
5. No term in the equation shows the variables as a product.

Not that you have to memorize any of this stuff—it's just that it's good to know what you can graph and what you can't graph.

# SOLVING FOR TWO VARIABLES

Let's review how to solve for two variables. We did it in Chapter 6, so it shouldn't be too much of a surprise for you. You can always go back to that chapter if you want to review the gory details of it all.

Now remember, we're going to choose a number to plug in for *x* so that we can find a pair to place with it as the *y* coordinate. Let's do this example:

$$4x + 2y = 10$$

Let's put in a 2 for the *x* coordinate and see what we get for *y*.

$$\begin{aligned} 4(\mathbf{2}) + 2y &= 10 \\ 8 + 2y &= 10 \\ 2y &= 10 - 8 \\ 2y = 2\ y &= \frac{2}{2} \\ y &= 1 \end{aligned}$$

So, one ordered pair for this equation would be (2, 1). Plug both the variables into the original equation to check your work:

$$\begin{aligned} 4x + 2y &= 10 \\ 4(\mathbf{2}) + 2(\mathbf{1}) &= 10 \\ 8 + 2 &= 10 \\ 10 &= 10 \end{aligned}$$

But remember, (2, 1) is only one possible answer. You can enter any number for *x* in order to get *y*. But if you plan to graph the coordinates, you might as well pick a nice easy number that's on your graph. That way, it will be easy to plot, right? Why make things hard on yourself?

# MORE POSSIBILITIES

Let's get a couple more coordinates that will fit into our equation. First, let's solve it with *x* as 1:

$$4x + 2y = 10$$
$$4(\mathbf{1}) + 2y = 10$$
$$4 + 2y = 10$$
$$2y = 10 - 4$$
$$2y = 6$$
$$y = \frac{6}{2}$$
$$y = 3$$

So, another coordinate pair is (1, 3). Don't forget to check it:

$$4(\mathbf{1}) + 2(\mathbf{3}) = 10$$
$$4 + 6 = 10$$
$$10 = 10$$

Let's do one more. How about *x* as 4:

$$4x + 2y = 10$$
$$4(\mathbf{4}) + 2y = 10$$
$$16 + 2y = 10$$
$$2y = 10 - 16$$
$$2y = -6$$
$$y = -\frac{6}{2}$$
$$y = -3$$

Our third ordered pair for this equation is (4, -3). Of course, give it a check by plugging in *x* and *y*:

$$4(\mathbf{4}) + 2(-\mathbf{3}) = 10$$
$$16 + (-6) = 10$$
$$16 - 6 = 10$$
$$10 = 10$$

# PLOTTING THE COORDINATES

Okay. We have three ordered pairs solved for one linear equation. That should give us plenty to work with when we go to plot points on our graph.

Now let's take our three coordinate pairs and plot them:

- (2, 1)
- (1, 3)
- (4, -3)

Then take some sort of straight edge—a ruler, a piece of paper, anything straight that you can use to make a straight line. Now connect all three plot points on your graph and make a straight line.

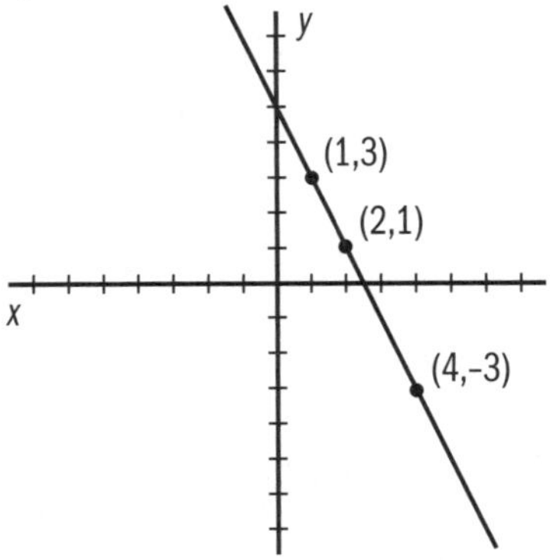


That line represents the linear equation $4x + 2y = 10$. Cool, huh?

# INTERCEPT

The intercept of a line is the point where the line you make crosses the *y*-axis. It's called the *y*-intercept. Take a look at the line you made. Where exactly does it cross the *y*-axis? It crosses the *y*-axis at 5, right? So the **y-intercept** of your line is 5.

And yes, you guessed it. The place where your line crosses the *x*-axis is called the ***x*-intercept**. In our example, the *x*-intercept is 2.5.

## DID YOU KNOW?

Here are some interesting facts about graphing. Look at your linear equation. If the coefficient of *x* is negative, your line will slant, or slope, to the left. If the coefficient of *x* is positive, the line will slope to the right. Two lines are parallel if they have the same slant, or slope.

The fastest way to convert a linear equation into a line on a graph is to assign *x* as 0 and solve for *y*. Then assign *y* as 0 and solve for *x*. You'll have two quick points to plot to make a line.

# YOUR TURN

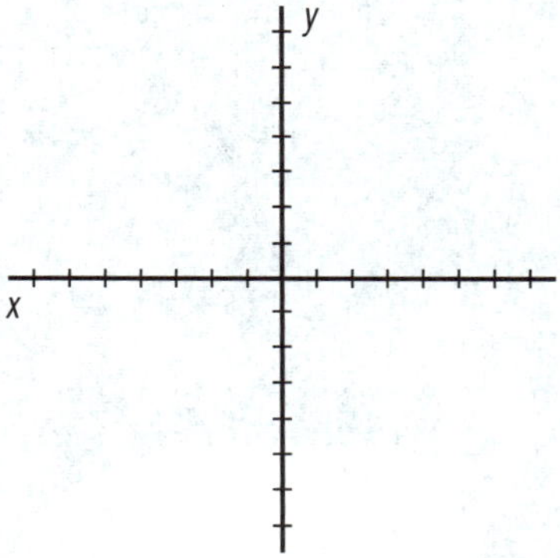


Use the following equation for questions 1, 2, and 3:

$$5x + 2y = 10$$

1. Write an ordered pair for this equation, where $x = 4$.
2. Find another ordered pair where $x = -2$.
3. Find another ordered pair where $y = 0$.

Use your ordered pairs to answer questions 4, 5, and 6.

4. Check and show your work for the ordered pair in question 1.
5. Check and show your work for the ordered pair in question 2.
6. Check and show your work for the ordered pair in question 3.

Use the above graph for questions 7 to 10.

7. Plot your three ordered pairs on the graph to make a line.
8. What is the $y$-intercept of your line?
9. What is the $x$-intercept of your line?
10. Does the ordered pair (4, 5) intercept the line that you drew? How can you tell without plotting it?

# 13

# CHAPTER 13
## TIPS

Now, I know you're not too interested in math. But it's never too late to learn.

"Ryan, let's get one thing straight. You're not an algebra tutor."

"Look, just because you don't like math doesn't mean you get to just skip it for the rest of your life."

"Ryan, I'm the school principal. I think I'd know if you were an authorized tutor."

"Algebra is my life."

"Ryan, you're not even good at math. I would never assign you as a tutor because you didn't even pass algebra. You had your accident before the final."

"I have algebra equations tattooed all over my body. Did you know that?"

"One of your crazy skater friends did that to you the day of your accident. You were planning to cheat on the final, apparently. Now, I have something to say to you that you're not going to like. I can't let you keep coming to school. Your memory isn't coming back, and you haven't made any progress in four years."

"I want to help people."

The principal sighs and shifts in his seat. "As soon as the board clears this decision, you will be removed from the school. I'm sorry, Ryan."

You know, if you want to read this book from cover to cover, that's just fine. But if you'd rather, you can look up key topics you're having trouble with and just shoot right to that section.

I'm Ryan. Algebra tutor *extraordinaire*.

People sometimes freak out when they see the letter *x*. I'm here to say, "Nah, it's not a problem."

# CHECKING YOUR ANSWERS

What's the most popular tip or word of advice from anyone when they talk to you about math? Check your work. Check your work. And check your work. You hear it so much, it's completely annoying.

Who wants to repeat a whole long problem again as soon as they finally get through it? You probably figure that if you did make any mistakes the first time around, you'll likely end up making those same mistakes the second time around. Or maybe you'll make new mistakes the second time.

Well, there is one really good thing about algebra in terms of checking your work. Once you find the value for your variable, you can go back and plug it into the original problem as a way to double-check your work. That way, you're not repeating the same steps. You're doing a new problem with new calculations, and you're checking your work at the same time.

One drawback to that, however, is that if you get the wrong answer, you can't find out what you did wrong. You still have to go back to the original problem and go over your work the old-fashioned way by repeating your steps and checking your calculations.

# SHOWING YOUR WORK

You've heard this one before too. Show your work. That's probably the second most annoying tip that people give you in math. Unfortunately, they are right.

Did you know that most standardized tests have multiple-choice answers? Great, you say. You won't have to show your work. You can just choose an answer from the list. Well, think again, my friend. Teachers sometimes want to see your work anyway. And it's a good idea too. That way, the teacher knows that you didn't cheat, that you didn't guess, and that you have some idea of how to solve the problem. Sometimes, partial credit can be given for setting up your problem and showing your work even if your calculations are incorrect. So, keep your records legible. It might help you in the long run.

# PRACTICE, PRACTICE, PRACTICE

This is another one of those annoying math tips you always hear: practice, practice, practice. Sorry again, but this one is true also.

Keep solving problems. Keep isolating variables. Keep reviewing what you don't know. Keep asking about what you don't understand. It's never too late to understand something. You can make a big difference on a test score or on a final grade if you keep trying. Nobody is born knowing all of these algebra rules. It's just not normal, and it's not common sense either. There are some really specific things that you have to practice and commit to memory so that you don't forget and so that you can pick up speed for tests.

# A FINAL FAREWELL

Well, I've said all that I want to say about algebra. I'm going to leave you with one final Your Turn section to chew on. You'll see a question or so from just about every chapter in the book. Go back and check out how to do a particular problem if you're stumped. Looking back in the book is definitely encouraged. Good luck with your algebra work.

## DON'T STRESS IT

Math is the number-one stress class for students in the United States. More students complain about not understanding math than about any other topic. And why not? It's really like speaking another language. Many number concepts are very abstract, and there are many symbols to learn. But it's all relative. Be glad you're not taking an advanced math class! As the brilliant mathematician and scientist Albert Einstein once said, "Do not worry about your difficulties in mathematics. I assure you that mine are greater."

# YOUR TURN

1. Find the absolute value:
   $|-22| + |14|$

2. Simplify the expression:
   $2\{x - 3[4 - (x + 1)] + 1\}$

3. Solve the equation:
   $(14n + 3n) + 2(5 + n) = 12 + 2n$

4. Solve the fractional expression:

   $$\frac{4 + 3x}{2} + 2x + \frac{5x - 1}{2}$$

5. Solve the equation for $y$ if $x = 2$:
   $4x + 14y = 12x + 2$

Solve the equation or expression:

6. $3(2x + 2^2) = 5x - 3^2$

7. $3^2(3^3)^4$

8. $(6x - 2)(x + 7)$

9. The sum of three consecutive numbers equals 45. Find those numbers.

10. Finish the ordered pair (2, ?) using this equation:
    $3x + 4y = 14$

TIPS

# ANSWER KEY

# CHAPTER 1

1. $y$
2. expression
3. 19
4. $12x$
5. $x + y$
6. $9 + 8x$
7. $8x - x$
8. $x - 1 + y$
9. $3x + 5y$
10. $450 \cdot 11 + 2x$
11. 4 times a number plus 0
12. 81 times a number minus another number
13. 2 times a number plus 3 times another number
14. 7 plus a number
15. 12 times a number minus 3 times another number

# CHAPTER 2

1. 2
2. 0
3. 7
4. $|142|$
5. 40
6. -4

   8 - 4 = 4. Make the 4 negative because the largest absolute value, 8, is negative.

7. 8

   14 - 6 = 8. Keep the 8 positive because the absolute value of 14 is positive.

8. -3

   Add the opposite of -2, which is 2. So, -5 + 2 = -3.

9. -15

Add the opposite of +7, which is -7. So, -8 + (-7) = -15.

10. 70

Add the opposite of -30, which is 30. So, 40 + 30 = 70.

11. 32

$8 \times 4 = 32$. Make the answer positive because the signs are same.

12. 42

$14 \times 3 = 42$. Make the answer positive because the signs are both negative.

13. -45

$5 \times 9 = 45$. Use the negative sign because the signs are different.

14. -165

$11 \times 15 = 165$. Use the negative sign because the signs are different.

15. -6

$12 \div 2 = 6$. Use the negative sign because the signs are different.

16. 5

$20 \div 4 = 5$. Keep the sign positive because both signs are same.

17. -17

9 + 8 = 17. Make the sign negative because both signs are negative.

18. -11

14 - 3 = 11. Make the sign negative because the 14 is negative.

19. -12

Add the opposite of 17, which is -17. 17 - 5 = 12. Keep the sign negative because the 17 is negative.

20. -40

$4 \times 10 = 40$. Make the sign negative because the signs are different.

# CHAPTER 3

1. $14x + 14y$

   $14(x + y)$

   $14x + 14y$

2. $17x + 8$

   $5(x - 4) + 4(3x + 7)$

   $5x - 20 + 12x + 28$

   $17x + 8$

3. $-5 + 3x$

   $x - [7 - 2(x + 1)]$

   $x - [7 - 2x - 2]$

   $x - 7 + 2x + 2$

   $x - 5 + 2x$

   $-5 + 3x$

4. $39x - 20$

   $5(7x - 3 + x) - (x + 5)$

   $35x - 15 + 5x - x - 5$

   $39x - 20$

5. $3 - 13x$

   $-3(4 + x) - 5[x - (3 - x)]$

   $-3(4 + x) - 5[x - 3 + x]$

   $-12 - 3x - 5x + 15 - 5x$

   $3 - 13x$

6. $10y + 3x$
$y(8 + 2) + x(4 - 1)$
$8y + 2y + 4x - x$
$10y + 3x$

7. $-6y^2 + 10y - 10x$
$3y(4 - 2y) - 2[y + (3x + 2x)]$
$3y(4 - 2y) - 2[y + 3x + 2x]$
$12y - 6y^2 - 2y - 6x - 4x$
$-6y^2 + 10y - 10x$

8. $8y - 20$
$2\{y - 3(6 - [2 + y]) + 3\} - 2$
$2\{y - 3(6 - 2 - y) + 3\} - 2$
$2\{y - 18 + 6 + 3y + 3\} - 2$
$2y - 36 + 12 + 6y + 6 - 2$
$8y - 20$

9. $-16 - 2x$
$4 + 2(x - 2[5 + x])$
$4 + 2(x - 10 - 2x)$
$4 + 2x - 20 - 4x$
$-16 - 2x$

10. $(65 + 4x)x$
$14(2x + 3x) - x(5 - 4x)$
$28x + 42x - 5x + 4x^2$
$65x + 4x^2$ or
$(65 + 4x)x$

# CHAPTER 4

1. $x = -2$

$(3x - 1) + 2 = (4x + 2) + 1$

$3x - 1 + 2 = 4x + 2 + 1$ Simplify.

$3x + 1 = 4x + 3$

$3x = 4x + 3 - 1$ Transpose.

$3x - 4x = 3 - 1$

$-x = 2$ Simplify.

$x = -2$ Divide by the coefficient.

2. $c = -2$

$-7 - 3c = -5 - 2c$

$-3c = -5 - 2c + 7$ Transpose.

$-3c = -2c + 2$

$-3c + 2c = 2$ Transpose.

$-1c = 2$ Simplify.

$$c = -\frac{2}{1}$$ Divide by the coefficient.

$c = -2$

3. $z = 3$

$-5(1 + z) + 12z = 16$

$-5(1) - 5(z) + 12z = 16$ Simplify.

$-5 - 5z + 12z = 16$

$7z = 16 + 5$ Transpose.

$7z = 21$ Simplify.

$$z = \frac{21}{7}$$ Divide by the coefficient.

$z = 3$

4. $w = 3$

$-3(6 - w) + 7w = 12$

$-3(6) - 3(-w) + 7w = 12$ Simplify.

$-18 + 3w + 7w = 12$

$-18 + 10w = 12$

$10w = 12 + 18$ Transpose.

$10w = 30$ Simplify.

$w = \frac{30}{10}$ Divide by the coefficient.

$w = 3$

5. $t = -\frac{3}{13}$

$-9 - (-3 - 5t) = -3(7t + 4)$

$-9 + 3 + 5t = -21t - 12$ Distribute on both sides.

$-6 + 5t = -21t - 12$ Simplify.

$5t = -21t - 12 + 6$ Transpose.

$5t = -21t - 6$ Transpose.

$21t + 5t = -6$ Simplify again.

$26t = -6$

$t = -\frac{6}{26}$ Divide by the coefficient.

$t = -\frac{3}{13}$ Simplify.

6. $x = \frac{3}{22}$

$14x + 6 - [(3 - x) - (7x + 9)] = 15$

$14x + 6 - [3 - x - 7x - 9] = 15$ Simplify.

$14x + 6 - [-6 - 8x] = 15$

$14x + 6 + 6 + 8x = 15$

$22x + 12 = 15$ Transpose.

$22x = 15 - 12$

$22x = 3$

Simplify.

$x = \frac{3}{22}$ Divide by the coefficient.

7. $x = -6$

$19 - \{[3x - (x - 1)] - 5x\} = 0$

$19 - \{[3x - x + 1] - 5x\} = 0$

$19 - \{2x + 1 - 5x\} = 0$

$19 - \{-3x + 1\} = 0$

$19 + 3x - 1 = 0$

$18 + 3x = 0$

$3x = -18$

$x = -\frac{18}{3}$

$x = -6$

8. $x = \frac{3}{7}$

$4[3 - (2x + 4x) + 3x] = 6 + 2x$

$4[3 - 2x - 4x + 3x] = 6 + 2x$

$4[3 - 3x] = 6 + 2x$

$12 - 12x = 6 + 2x$

$-12x - 2x = 6 - 12$

$-14x = -6$

$14x = 6$

$x = \frac{6}{14}$

$$x = \frac{3}{7}$$

9. $y = \frac{5}{8}$

$(4y + 7) + (3 - 2y) = 18y$

$4y + 7 + 3 - 2y = 18y$

$2y + 10 = 18y$

$10 = 18y - 2y$

$10 = 16y$

$$\frac{10}{16} = y$$

$$\frac{5}{8} = y$$

10. $x = \frac{10}{11}$

$3(4x - 2) + (3 - x) = 7$

$12x - 6 + 3 - x = 7$

$11x - 3 = 7$

$11x = 7 + 3$

$11x = 10$

$$x = \frac{10}{11}$$

# CHAPTER 5

1. $x = \frac{1}{2}$

$$x - \frac{x+1}{2} = 4x - (1 + 2x) - \frac{x}{2}$$

$$x - \frac{x+1}{2} = 4x - 1 - 2x - \frac{x}{2}$$

$$2\left(x - \frac{x+1}{2} = 4x - 1 - 2x - \frac{x}{2}\right)$$

$$2x - \frac{\cancel{2}(x+1)}{\cancel{2}} = 8x - 2 - 4x - \frac{\cancel{2}x}{\cancel{2}}$$

$2x - x - 1 = 3x - 2$

$x - 1 = 3x - 2$

$x - 3x = -2 + 1$

$-2x = -1$

$$x = \frac{1}{2}$$

2. $$x = \frac{55}{4}$$

$$10 - \frac{3x}{5} = x - 12$$

$$5\left(10 - \frac{3x}{5} = x - 12\right)$$

$$50 - \frac{\cancel{5}(3x)}{\cancel{5}} = 5x - 60$$

$50 - 3x = 5x - 60$

$50 + 60 = 5x + 3x$

$110 = 8x$

$$\frac{110}{8} = x$$

$$\frac{55}{4} = x$$

3. $x = 5$

$$2(x-2) = 5 + \frac{x-1}{4}$$

$$2x - 4 = 5 + \frac{x-1}{4}$$

$$4\left(2x - 4 = 5 + \frac{x-1}{4}\right)$$

$$8x - 16 = 20 + \frac{\cancel{4}(x-1)}{\cancel{4}}$$

$8x - 16 = 20 + x - 1$
$8x - 16 = 19 + x$
$8x - x = 19 + 16$
$7x = 35$

$$x = \frac{35}{7}$$

$x = 5$

4. $x = 2$

$$\frac{5}{x} = \frac{1}{2} + \frac{2 + x}{x}$$

$$x\left(\frac{5}{x} = \frac{1}{2} + \frac{2 + x}{x}\right)$$

$$2x\left(\frac{5}{x}\right) = 2x\left(\frac{1}{2}\right) + 2x\left(\frac{2 + x}{x}\right)$$

$$2\cancel{x}\left(\frac{5}{\cancel{x}}\right) = \cancel{2}x\left(\frac{1}{\cancel{2}}\right) + 2\cancel{x}\left(\frac{2 + x}{\cancel{x}}\right)$$

$10 = x + 2(2 + x)$
$10 = x + 4 + 2x$
$10 = 3x + 4$
$10 - 4 = 3x$
$6 = 3x$

$$\frac{6}{3} = x$$

$2 = x$

5. $5 = 5$

$$\frac{5}{x} = \frac{1}{2} + \frac{2 + x}{x}$$

$$\frac{5}{2} = \frac{1}{2} + \frac{2 + 2}{2}$$

$$\frac{5}{2} = \frac{1}{2} + \frac{4}{2}$$

$$2\left(\frac{5}{2} = \frac{1}{2} + \frac{4}{2}\right)$$

$$\cancel{2}\left(\frac{5}{\cancel{2}} = \frac{1}{\cancel{2}} + \frac{4}{\cancel{2}}\right)$$

5 = 1 + 4
5 = 5

6. $x = -7$

$$13 + \frac{3}{x} = 12 - \frac{4}{x}$$

$$x\left(13 + \frac{3}{x} = 12 - \frac{4}{x}\right)$$

$$13x + \frac{\cancel{x}(3)}{\cancel{x}} = 12x - \frac{\cancel{x}(4)}{\cancel{x}}$$

$13x + 3 = 12x - 4$
$13x - 12x = -4 - 3$
$x = -7$

7. -88 = -88

$$13 + \frac{3}{x} = 12 - \frac{4}{x}$$

$$13 + \frac{3}{-7} = 12 - \frac{4}{-7}$$

$$-7\left(13 + \frac{3}{-7} = 12 - \frac{4}{-7}\right)$$

$$-91 + \frac{\cancel{-7}(3)}{\cancel{-7}} = -84 - \frac{\cancel{-7}(4)}{\cancel{-7}}$$

-91 + 3 = -84 - 4
-88 = -88

8. $x = \frac{22}{3}$

$$\frac{8+3}{x} = \frac{5-2}{2}$$

$2(8 + 3) = x(5 - 2)$
$16 + 6 = 5x - 2x$
$22 = 3x$
$\frac{22}{3} = x$

9. $x = 135$

$$\frac{15}{8x} = \frac{2}{9+x}$$

$15(9 + x) = 2(8x)$
$135 + 15x = 16x$
$135 = 16x - 15x$
$135 = x$

10. $r = -\frac{22}{7}$

$$\frac{26}{8-r} = \frac{2}{4+r}$$

$26(4 + r) = 2(8 - r)$
$104 + 26r = 16 - 2r$
$104 - 16 = -2r - 26r$
$88 = -28r$

$-\frac{88}{28} = r$

$-\frac{22}{7} = r$

# CHAPTER 6

1. $x = -\frac{7}{8}y$

$13x + 4y = 5x - 3y$

$13x - 5x = -3y - 4y$

$8x = -7y$

$x = -\frac{7}{8}y$

2. $n = 3$

$6r + 4n = 12$ $(r = 0)$

$6(0) + 4n = 12$

$4n = 12$

$n = \frac{12}{4}$

$n = 3$

3. **(3, 0)**

$3(3) + 2y = 9$

$9 + 2y = 9$

$2y = 9 - 9$

$2y = 0$

$y = 0$

4. **(–3, 9)**

$3(-3) + 2y = 9$

$-9 + 2y = 9$

$2y = 9 + 9$

$2y = 18$

$y = \frac{18}{2}$

$y = 9$

5. **(3, 0)**
$3x + 2(0) = 9$
$3x + 0 = 9$
$3x = 9$
$$x = \frac{9}{3}$$
$x = 3$

6. **(–5, 12)**
$3x + 2(12) = 9$
$3x + 24 = 9$
$3x = -15$
$$x = -\frac{15}{3}$$
$x = -5$

7. yes
$2x + 3y = 12$ (3, 2)
$2(3) + 3(2) = 12$
$6 + 6 = 12$

8. no
$y = 3x + 1$ (10, 3)
$3 = 3(10) + 1$
$3 = 30 + 1$
$3 \neq 31$

9. yes
$2a - 3b = 20$ (16, 4)
$2(16) - 3(4) = 20$
$32 - 12 = 20$
$20 = 20$

10. no

$8g - 3h = 8$ (3, 5)

$8(3) - 3(5) = 8$

$24 - 15 = 8$

$9 \neq 8$

# CHAPTER 7

1. 89

$(3 \times 3 \times 3 \times 3) + (2 \times 2 \times 2)$

$3^4 + 2^3 = 89$

$81 + 8 = 89$

2. 7,221

$5^5 + 8^4 =$

$(5 \times 5 \times 5 \times 5 \times 5) + (8 \times 8 \times 8 \times 8) =$

$3{,}125 + 4{,}096 = 7{,}221$

3. $x = -4\frac{1}{6}$

$4(2x + 2^2) + 4x = 6x - 3^2$

$4(2x + 4) + 4x = 6x - 3^2$ Work within parentheses first.

$8x + 16 + 4x = 6x - 9$ Simplify any other exponents.

$8x + 4x - 6x = -9 - 16$

$6x = -25$

$x = -\frac{25}{6}$

$x = -4\frac{1}{6}$

4. 4.

$(8 + 3^3) + (4 - 5^4) = x + 20$

$8 + 27 + 4 - 625 = x + 20$

$-586 = x + 20$

$-586 - 20 = x$

$-606 = x$

5. 5.

$x(3^2 + 2^3) \Diamond 3x(5^2) = 50$

$x(9 + 8) \Diamond 3x(25) = 50$

$x(17) \Diamond 3x(25) = 50$

$17x \Diamond 75x = 50$

$1{,}275x^2 = 50$

$$x^2 = \frac{2}{51}$$

$$x = \sqrt{\frac{2}{51}}$$

6. 6.

$\sqrt{81} + \sqrt{169}$

$9 + 13 = 22$

7. 7.

$\sqrt{147}$

$\sqrt{49 \times 3}$

$\sqrt{49} \times \sqrt{3}$

$7\sqrt{3}$

8. 8.

$5\sqrt{12} - 13\sqrt{27} + 17\sqrt{45}$

$5\sqrt{4 \times 3} - 13\sqrt{9 \times 3} + 17\sqrt{9 \times 5}$ Simplify roots.

$5(2)\sqrt{3} - 13(3)\sqrt{3} + 17(3)\sqrt{5}$ Remove perfect squares.

$10\sqrt{3} - 39\sqrt{3} + 51\sqrt{5}$

$-29\sqrt{3} + 51\sqrt{5}$

9. 9.

$\sqrt{5}(2 - \sqrt{3})$

$2\sqrt{5} - \sqrt{15}$

10. 10.

$-2\sqrt{12} \times (-5\sqrt{3})$

$10\sqrt{36}$ Multiply coefficients and roots.

$10 \times 6 = 60$

# CHAPTER 8

1. $4 \cdot 4^5 = 4^6$ can be solved with the multiplication law.
2. $(2^3)^2$ can be solved with the power of a power law.
3. $(5 \cdot 7)^3 = 5^3 \cdot 7^3$ can be solved with the power of a product law.
4. 5

$5x^0$

$5 \cdot 1 = 5$

5. $x^{22}$

$x^2(x^4)^5$

$x^2 \cdot x^{20} = x^{20+2} = x^{22}$

6. $256x^{10}$

$(4x^2)^3(2x^2)^2$

$4^3x^6 \cdot 2^2x^4$

$64x^6 \cdot 4x^4$

$256x^{10}$

7. 7.

$(x^3y)(xy^2)$

$x^{3+1} \cdot y^{1+2}$

$x^4y^3$

8. $\frac{16}{729}$

$\left(\frac{2^2}{3^3}\right)^2$

$\frac{2^{2 \cdot 2}}{3^{3 \cdot 2}} = \frac{2^4}{3^6} = \frac{16}{729}$

9. 25

$\frac{5^4}{5^2}$

$5^{4-2} = 5^2 = 25$

10. $\frac{4y^2}{9x^4}$

$\frac{(2xy)^2}{(3x^2)^3}$

$\frac{2^2x^2y^2}{3^3x^6} = \frac{4x^2y^2}{9x^6} = \frac{4y^2}{9x^4}$

# CHAPTER 9

1. $6x^2 + 56x + 18$

$(6x + 2)(x + 9)$

**F**irst $6x \cdot x =$ **6x2**

**O**uter $6x \cdot 9 =$ **54x**

**I**nner $2 \cdot x =$ **2x**

**L**ast $2 \cdot 9 =$ **18**

$6x^2 + 54x + 2x + 18$

$6x^2 + 56x + 18$

2. $4x^2 - x - 3$

$(4x + 3)(x - 1)$

First $4x \cdot x =$ **4x2**

Outer $4x \cdot (-1) =$ **$-4x$**

Inner $3 \cdot x =$ **3x**

Last $3 \cdot (-1) =$ **$-3$**

$4x^2 - 4x + 3x - 3$

$4x^2 - x - 3$

3. $15b^2 - b - 6$

$(5b + 3)(3b - 2)$

First $5b \cdot 3b =$ **15b2**

Outer $5b \cdot (-2) =$ **$-10b$**

Inner $3 \cdot 3b =$ **9b**

Last $3 \cdot (-2) =$ **$-6$**

$15b^2 - 10b + 9b - 6$

$15b^2 - b - 6$

4. $36z^2 - 16$

$(6z - 4)(6z + 4)$

First $6z \cdot 6z =$ **36z2**

Outer $6z \cdot 4 =$ **24z**

Inner $-4 \cdot 6z =$ **$-24z$**

Last $-4 \cdot 4 =$ **$-16$**

$36z^2 + 24z - 24z - 16$

$36z^2 - 16$

5. $36x^2y^2 + 78xy - 14$

$(12xy - 2)(3xy + 7)$

First $12xy \cdot 3xy =$ **36x2y2**

Outer $12xy \cdot 7 =$ **84xy**

Inner $-2 \cdot 3xy =$ **$-6xy$**

Last $-2 \cdot 7 =$ **$-14$**

$36x^2y^2 + 84xy - 6xy - 14$

$36x^2y^2 + 78xy - 14$

6. $25y^2 + 70y + 49$
$(5y + 7)^2$
To square a binomial, use the formula
$(a - b)^2 = a^2 - 2ab + b^2$
$(5y)^2 + 2(5y)(7) + 7^2$
$25y^2 + 70y + 49$

7. $64z^2 - 80z + 25$
$(8z - 5)^2$
To square a binomial, use the formula
$(a - b)^2 = a^2 - 2ab + b^2$
$(8z)^2 - 2(8z)(5) + (5)^2$
$64z^2 - 80z + 25$

8. $x^3 - 2x^2 - x + 2$
$(x + 1)(x^2 - 3x + 2)$
$x^3 - 3x^2 + 2x + x^2 - 3x + 2$
$x^3 - 2x^2 - x + 2$

9. $6x^3 + 7x^2 - 3x - 1$
$(3x^2 + 5x + 1)(2x - 1)$
To make distributing easier, rewrite as
$(2x - 1)(3x^2 + 5x + 1)$
$6x^3 + 10x^2 + 2x - 3x^2 - 5x - 1$
$6x^3 + 7x^2 - 3x - 1$

10. $3y^3 + y^2 - 2y - 2$
$(y - 1)(3y^2 + 4y + 2)$
$3y^3 + 4y^2 + 2y - 3y^2 - 4y - 2$
$3y^3 + y^2 - 2y - 2$

# CHAPTER 10

1. If $n$ is the first number, then the second number would be $\frac{4}{5}n$.

You can put any numbers in for $n$. For example if $n = 5$, you could find the second number by writing $\frac{4}{5} \times 5 = 4$.

**Original question:** One number is $\frac{4}{5}$ of another.

2. Let $x$ equal Sara's age. If Margo is three times older than Sara, Margo's age is $3x$.
**Original question:** Margo is three times older than Sara.

3. Let $m$ equal the number of quarters Mike has. That means Jeff has $8m$ quarters. For example, if Mike has 2 quarters, Jeff has 16. If Mike has 5 quarters, Jeff has 40.
**Original question:** Jeff has eight times as many quarters as Mike.

4. If we call the first number $n$, then the next consecutive integer would be $n + 1$. So, if the first integer is 10, the second is 10 + 1, or 11.
**Original question:** Two numbers are consecutive integers.

5. If $l$ is the length of the rectangle, the width is $7 + 3l$. So, if 10 is the length, the width is 7 + 3(10), or 37.
**Original question:** The width of a rectangle is seven more than three times its length.

6. Let $n$ equal "a number." Then, another number is $5n - 9$. So, if one number is 2, the other is 5(2) – 9, or 1.
**Original question:** Five times a number decreased by nine is equal to another number.

7. Let $n$ equal the smaller integer. Then, $n + 1$ is the next consecutive integer.

$n + (n + 1) = 39$

$2n + 1 = 39$

$2n = 39 - 1$

$2n = 38$

$$n = \frac{38}{2}$$

$n = 19, n + 1 = 20$

**Original question:** The sum of two consecutive integers is 39. Find the integers.

8. Let $n$ = the number.

$3n - 9 = 3 + 2n$

$3n - 2n = 3 + 9$

$n = 12$

Check the answer:

$3(12) - 9 = 3 + 2(12)$

$36 - 9 = 3 + 24$

$27 = 27$

It checks out.

**Original question:** Nine less than three times a number is equal to three more than twice the number. Find the number.

9. Let $k$ = Kate's age. Let $6k$ = Laura's age.

$k + 6k = 28$

$7k = 28$

$$k = \frac{28}{7}$$

$k = 4$

Since $6k$ = Laura's age, Laura's age = 6(4) = 24.

*Check the answer:*

$4 + 6(4) = 28$

$4 + 24 = 28$

$28 = 28$

**Original question:** Laura is six times older than Kate. The sum of their ages is 28. How old are Laura and Kate?

10. Let $n$ be the smallest of the consecutive integers.
Then, $n + 1$ = the middle integer, and $n + 2$ = the largest integer.

$$n + (n + 1) - 11 = n + 2$$
$$2n + 1 - 11 = n + 2$$
$$2n - 10 = n + 2$$
$$2n - n = 2 + 10$$
$$n = 12$$

If the smallest integer is 12, the middle one is 13, and the largest is 14.
**Original question:** Find three consecutive integers such that the sum of the first two decreased by 11 is equal to the third.

# CHAPTER 11

1. Quadrant II
2. Quadrant I
3. Quadrant IV
4. Quadrant III

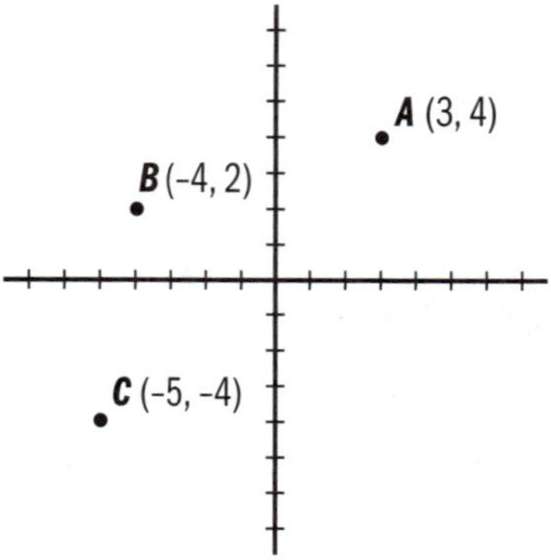


5. $A$ (3, 4)
6. $B$ (-4, 2)
7. $C$ (-5, -4)
8. (-5, 4)
9. (0, -3)
10. $\left(\frac{1}{2}, -2\right)$

# CHAPTER 12

1. (4, -5)

   $5x + 2y = 10$

   $5(4) + 2y = 10$

   $20 + 2y = 10$

   $2y = 10 - 20$

   $2y = -10$

   $y = -\frac{10}{2}$

   $y = -5$

   The ordered pair is (4, -5).

2. (-2, 10)

   $5x + 2y = 10$

   $5(-2) + 2y = 10$

   $-10 + 2y = 10$

   $2y = 10 + 10$

   $2y = 20$

   $y = \frac{20}{2}$

   $y = 10$

   The ordered pair is (-2, 10).

3. (2, 0)

   $5x + 2y = 10$

   $5x + 2(0) = 10$

   $5x + 0 = 10$

   $5x = 10$

   $x = \frac{10}{5}$

   $x = 2$

   The ordered pair is (2, 0).

4. $10 = 10$

$5x + 2y = 10$

$5(4) + 2(-5) = 10$

$20 - 10 = 10$

$10 = 10$

5. $10 = 10$

$5x + 2y = 10$

$5(-2) + 2(10) = 10$

$-10 + 20 = 10$

$10 = 10$

6. $10 = 10$

$5x + 2y = 10$

$5(2) + 2(0) = 10$

$10 + 0 = 10$

$10 = 10$

7.

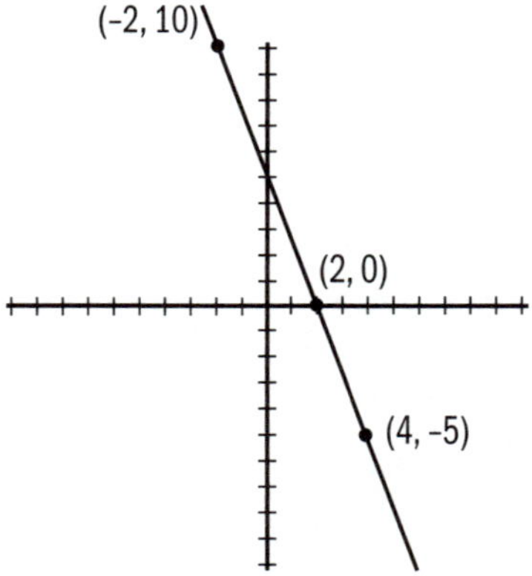


8. The $y$-intercept is 5.
9. The $x$-intercept is 2.

10. No. The coordinates don't make the original equation true:

$5x + 2y = 10$

$5(4) + 2(5) \neq 10$

$20 + 10 \neq 10$

$30 \neq 10$

# CHAPTER 13

1. 36

$|-22| + |14|$

$22 + 14 = 36$

2. $8x - 16$

$2\{x - 3[4 - (x + 1)] + 1\}$

$2\{x - 3[4 - x - 1] + 1\}$

$2\{x - 3[3 - x] + 1\}$

$2\{x - 9 + 3x + 1\}$

$2\{4x - 8\}$

$8x - 16$

3. $\frac{2}{17}$

$(14n + 3n) + 2(5 + n) = 12 + 2n$

$14n + 3n + 10 + 2n = 12 + 2n$

$19n + 10 = 12 + 2n$

$19n - 2n = 12 - 10$

$17n = 2$

$n = \frac{2}{17}$

4. $3 + 12x$

$\frac{4 + 3x}{2} + 2x + \frac{5x - 1}{2}$

$2\left(\frac{4 + 3x}{2} + 2x + \frac{5x - 1}{2}\right)$

$$\frac{\not{2}(4 + 3x)}{\not{2}} + 4x + \frac{\not{2}(5x - 1)}{\not{2}}$$

$4 + 3x + 4x + 5x - 1$

$3 + 12x$

5. $y = 1\frac{2}{7}$

$4x + 14y = 12x + 2$

$4(2) + 14y = 12(2) + 2$

$8 + 14y = 24 + 2$

$14y = 24 + 2 - 8$

$14y = 18$

$y = \frac{18}{14}$

$y = 1\frac{2}{7}$

6. $x = -21$

$3(2x + 2^2) = 5x - 3^2$

$3(2x + 4) = 5x - 9$

$6x + 12 = 5x - 9$

$6x - 5x = -9 - 12$

$x = -21$

7. $3^{14} = 4{,}782{,}969$

$3^2(3^3)^4$

$3^2 \cdot 3^{12} = 3^{2 + 12} = 3^{14}$

$3^{14} = 4{,}782{,}969$

8. $6x^2 + 40x - 14$

$(6x - 2)(x + 7)$

$6x^2 + 42x - 2x - 14$

$6x^2 + 40x - 14$

9. 14, 15, 16

$n$ = the first number

$n + 1$ = the second number

$n + 2$ = the third number

$n + (n + 1) + (n + 2) = 45$

$3n + 3 = 45$

$3n = 45 - 3$

$3n = 42$

$$n = \frac{42}{3}$$

$n = 14$

$14 + 15 + 16 = 45$

10. $y = 2$

$3x + 4y = 14$ (2, ?)

$3(2) + 4y = 14$

$6 + 4y = 14$

$4y = 14 - 6$

$4y = 8$

$$y = \frac{8}{4}$$

$y = 2$